THE ECONOMIES OF WEST AFRICA

THE INTERNATIONAL ECONOMIES SERIES

ALREADY PUBLISHED:
The Japanese Economy by G. C. Allen

FORTHCOMING:
The Economy of India by V. N. Balasubramanyam

THE ECONOMIES OF WEST AFRICA

Douglas Rimmer
Director, Centre of West African Studies,
University of Birmingham

WEIDENFELD AND NICOLSON
LONDON

In memory of
June

British Library Cataloguing in Publication Data

Rimmer, Douglas
The economies of West Africa.—(The International economies series; 2)
1. Africa, West – Economic conditions
I. Title II. Series
330.966 HC17.W.5

ISBN 0 297 78095 6 cased
ISBN 0 297 78096 4 paperback

George Weidenfeld & Nicolson Limited
91 Clapham High Street, London SW4 7TA

Filmset by Deltatype, Ellesmere Port
Printed and bound in Great Britain by
Butler and Tanner Ltd
Frome & London

CONTENTS

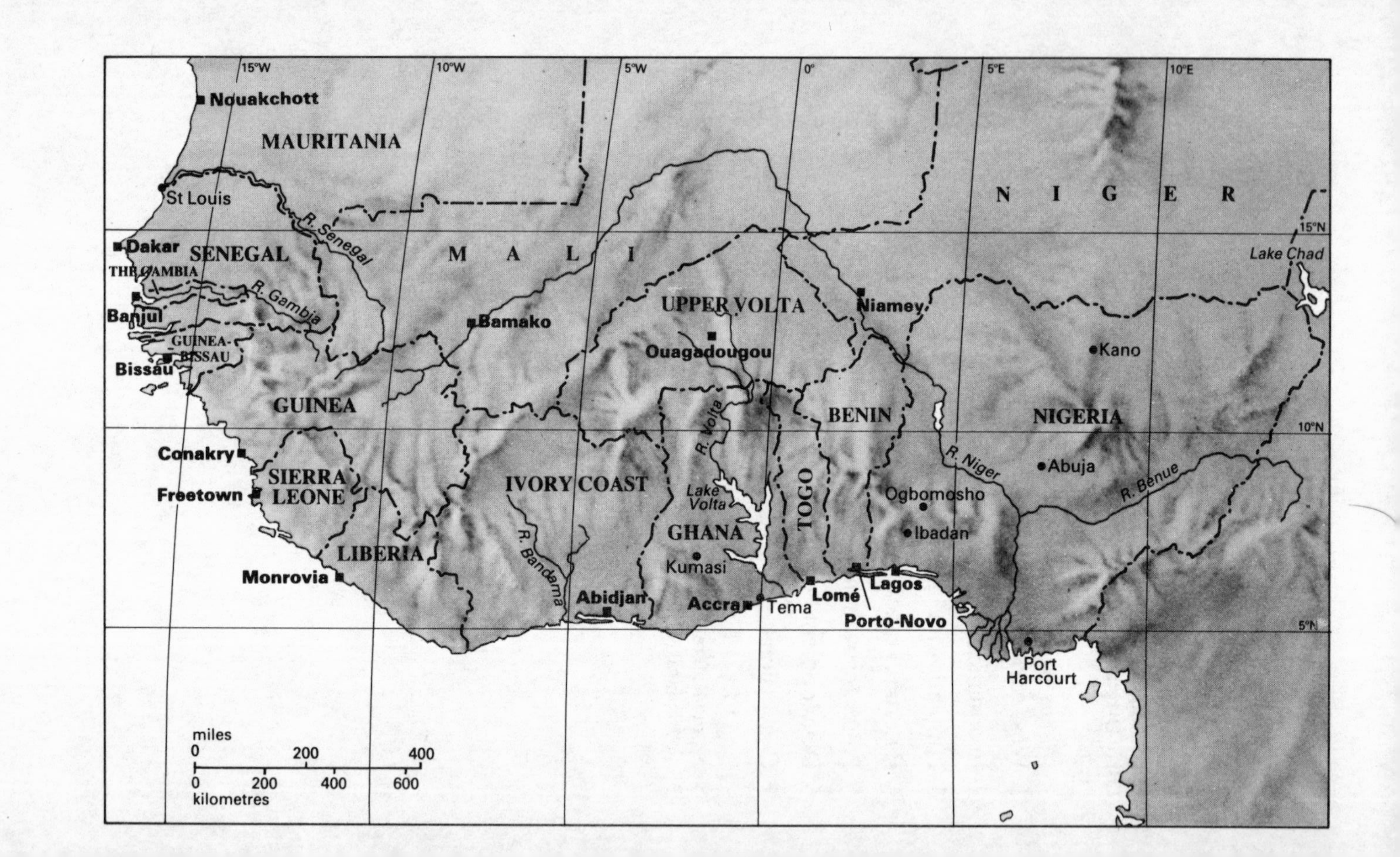
15°W
10°W
5°W
0°
5°E
10°E
15°N
10°N
5°N
Nouakchott
MAURITANIA
St Louis
R. Senegal
NIGER
Dakar
SENEGAL
THE GAMBIA
R. Gambia
Banjul
GUINEA-BISSAU
Bissau
MALI
Bamako
UPPER VOLTA
Niamey
Ouagadougou
Lake Chad
Kano
GUINEA
BENIN
NIGERIA
Conakry
SIERRA LEONE
Freetown
IVORY COAST
R. Volta
Lake Volta
TOGO
R. Niger
Abuja
R. Benue
Ogbomosho
Ibadan
GHANA
Kumasi
LIBERIA
R. Bandama
Monrovia
Abidjan
Accra
Tema
Lomé
Lagos
Porto-Novo
Port Harcourt
miles
0
200
400
0
200
400
600
kilometres

TABLES

DIAGRAMS

PREFACE

This book is intended for students of economics who wish to be informed about West Africa, and for students of West Africa who require some understanding of the economic facts and relationships relevant to that area. The former may find salutary the emphases which are placed on the unreliability of West African statistics and on the inevitably political character of economic policies. The latter may benefit from an appreciation of relative economic magnitudes, however imperfect the measurement of those quantities may be, and they may be persuaded that economic policies have economic consequences which are not necessarily those intended or said to be intended. The economic reasoning which I have used (theorizing seems too grand a word) is of an elementary kind, and it is hoped that the book will be widely accessible.

West Africa has been defined as the area of ECOWAS – the Economic Community of West African States. Cameroon is therefore omitted. Its inclusion would, I think, have strengthened the conclusions which emerge from the book and are encapsulated in the final chapter. ECOWAS consists of sixteen sovereign entities. They are not of equal significance economically or in population, and they have not been treated equally in this book. While the Cape Verde Republic and Guinea-Bissau are the only countries on which I have found extremely little to say, I have concentrated among the rest on Nigeria, Ghana and the Ivory Coast – the three countries which were estimated (unreliably, of course) to account in 1979 for 70 per cent of West Africa's population and 85 per cent of its economic production.

After an introductory chapter, I explore the economic anatomy of the region under the headings of economic structures, population and labour force, and external trade. Wherever possible, I cite comparable economic and demographic estimates for the several countries, and I essay measurements for West Africa considered as a whole. I am concerned with changes over time as well as with the phenomena currently or recently apparent. Often I take 1960, the year of political independence for most West African states, as a benchmark, but it has also been useful or necessary to make references further back in time, since certain sustained economic trends had their beginnings, and some important economic institutions were created, in the final phase of colonialism.

In Chapters 5 and 6, I turn to physiology, or the functioning of the West African economies – and also to pathology, or their malfunctioning. I make policy pivotal in the exposition, partly because government action in the economy is practically important from any standpoint, and partly because it has long been claimed and believed to be the fulcrum on which the material conditions of life in West Africa could be raised. Chapter 5 is concerned with the instruments of policy and their imperfections. In Chapter 6 I review the development objectives which have been pursued or professed and the development strategies which have been in vogue. As to the latter, I identify two central propositions – that the rate of investment should be compulsorily raised and the composition of investment administratively controlled, and that economic 'self-reliance' should be contrived and fostered on a national (or, occasionally, a regional or continental) basis. I find these propositions implausible as statements of conditions of economic progress, and suggest that their acceptance tends rather toward impoverishment.

The final chapter draws together my conclusions. They are that rapid economic growth, and considerable social amelioration, were achieved in West Africa in the twenty years following 1960 (and in the thirty-five years following the Second World War). Policies have been substantially, albeit only roughly, congruent with demographic and economic advancement so far as concerns the uses to which economic gains have been put. Policies have been much less effective in the active winning of those gains. Often, their contribution in that direction has been negative. The principal source of the achievements has been the response of West African resources to market opportunities overseas. Foreign enterprise and aid have also been important. It follows that the achievements have been less in areas which are naturally disadvantaged, as is the Sahel, and in countries like Ghana and Guinea, where governments have been more than usually hostile to external economic relationships.

While no great novelty could be claimed for these conclusions, I am conscious that they are unlikely to be found generally palatable. In the hope, no doubt forlorn, of avoiding misunderstanding, let me say now that I do not question that West African countries ought to become further industrialized, economically diversified and modernized. I question only the wisdom of attempting these changes prematurely, or when they can be accomplished only with grievous inefficiency, or at unreasonably high costs to residents of the countries concerned. I suggest that those who make light of these costs hold too small the values and interests of their fellow-citizens. In the long run, of course, even the most horrendous economic distortions may appear as if they had been justified by later developments, but in the meantime people will have suffered, and, as Brecht observed, nothing you can do will help a dead man.

For what I hope will be the convenience of readers, I have summarized each chapter except the last in its concluding section. Unfamiliar acronyms are explained in the index.

It remains to express my thanks to my publishers, for their patience in awaiting completion of this work; to Peter Mitchell and Margaret Peil, for relieving my ignorance of some matters; to Kim Davies and Vivien Wright, for secretarial assistance; and to Nellie Bennett, Marion Johnson and, above all, Joyce Rimmer, for the constant encouragement they have given me.

Douglas Rimmer
June 1983

1
West Africa as an Economic Region

The delimitation of West Africa

West Africa is conventionally delimited as the area bounded by the Atlantic and lying south of the Sahara and west of Cameroon. The geographical individuality of this area is debatable. Its intercourse with North Africa has been sustained over many centuries, the desert notwithstanding, and for long was less difficult than its seaborne traffic. On its eastern boundary, the mountains of Cameroon and Adamawa are a limited barrier; north of them, between the forest and the desert, an open plain stretches across the continent from the Atlantic to the Red Sea. The ethnographical distinctiveness of West Africa may also be questioned. To the Arabs of North Africa, it was *Bilad-es-Sudan*, the land of the negroes, but this was an appellation that could have been given to all sub-Saharan Africa outside Ethiopia and the Horn and the extreme southwest. Linguists make a distinction between the indigenous languages of West Africa and the Bantu tongues of other negroes, but its significance is uncertain.

An alternative conception of West Africa is possible; on several grounds, including cultural and botanical differences as well as relief, a division might be made between the highlands of East and Central Africa and a generally less elevated region which would include much of the Zaïre (or Congo) and Chad basins as well as the lands west of Cameroon.[1] Such a western Africa would include all the major areas of African rain forest as well as the savanna further north and would roughly coincide with the area in which European settlement was slight. Its seaboard would correspond with what the British originally understood by the Guinea Coast, or simply 'the Coast'.

That West Africa as conventionally defined is more closely circumscribed owes more to diplomacy and administration than to geography or ethnography. In the last twenty years of the nineteenth century, the French pursued a campaign of territorial aggrandizement from their long-established base on the Senegal river and trading posts on the Ivory, Slave and Gabon coasts. The eventual result was an empire of contiguous territory stretching from the Mediterranean coast to the Congo river and from Cap Vert to Darfur. Other European powers with trading interests or territorial ambitions in Africa asserted rights of dominion or protection against this envelopment – the British in Nigeria, the Gold Coast, Sierra Leone and The Gambia; the Germans in Cameroon and Togoland; King Leopold of the Belgians in the Congo; the Portuguese in Angola and some smaller territories. So the map of colonial western Africa was drawn. One independent state, Liberia, survived, though its rulers were not natives of the country but descendants of settlers brought across the Atlantic.

Within the vast French dominion, the administration of the more westerly territories was organized separately from that of the more southerly, and also from Algeria. In 1895, the governor at St Louis in Senegal became also a governor-general with responsibilities extending as far as Niger and the Ivory Coast. Dahomey was added later, and in 1904 the government-general of *Afrique occidentale française* (AOF) was moved to Dakar and detached from the administration of Senegal. The eventual constituents of this colonial federation were to be the territories of Dahomey, Guinea, the Ivory Coast, Mauritania, Niger, Senegal, Soudan and Upper Volta. A little later, the more southerly lands (Gabon, Middle Congo, and Oubangui-Chari-Chad) were similarly constituted as *Afrique équatoriale française* (AEF).

Thus it was French colonialism that drew the distinction between West and Equatorial Africa. The four British territories that have been mentioned were not federated, but they shared some joint services and it became common to refer to them collectively as British West Africa in contradistinction from AOF or French West Africa. Cameroon, the hinge between West and Equatorial Africa, was lost by the Germans in 1916 and later mandated by the League of Nations to Great Britain and France. The British trust, known as the Northern and Southern Cameroons, was adjacent to Nigeria and administered along with that country, while the French mandate, though administered separately, was contiguous to AEF. When these trust territories became politically independent in 1960, the new Federal Republic of Cameroon (comprising the former French territory and the Southern Cameroons)[2] sought to develop economic cooperation with the former constituents of AEF rather than with its westerly neighbours.

The countries of British West Africa and of AOF (along with the French trust territory of Togo) all attained independence between 1957 and 1965, and such elements of economic cooperation as had existed within each of these two groups – and which in the latter case were to some extent maintained after independence – encouraged the United Nations Economic Commission for Africa in the mid-1960s to identify West Africa as a potential arena of international economic integration. Largely on the initiative of Nigeria, a treaty to establish an Economic Community of West African States (ECOWAS) was in fact made in 1975 by all fifteen mainland states composing the West Africa of conventional definition. The Cape Verde islands became the sixteenth member of this Community in 1977.

Such are the ways in which the notional boundaries of West Africa (and the area with which this book is concerned) have become established. The sixteen West African states and their dates of independence are as follows:

Former British West Africa
Ghana (formerly the Gold Coast), 1957
Nigeria, 1960
Sierra Leone, 1961
The Gambia, 1965

Former AOF
Guinea, 1958
Benin (formerly Dahomey), 1960
Ivory Coast, 1960
Mali (formerly Soudan), 1960
Mauritania, 1960
Niger, 1960
Senegal, 1960
Upper Volta, 1960

Others
Liberia, sovereignty proclaimed in 1847
Togo, 1960
Guinea-Bissau (formerly Portuguese Guinea), 1974
Cape Verde, 1975

Climatic conditions

Together the sixteen states occupy nearly 2.5 million square miles (6.5 million square kilometres), an area larger than Europe excluding Russia. Nearly half of this area is Saharan desert enclosed within the borders of Mauritania, Mali and Niger. The remainder extends, on the mainland, roughly 800 miles (1,300 km) between about 5° and 18°N and 2,000 miles (3,200 km) from west to east. The amount and incidence of rainfall are the chief determinants of the extent and character of vegetation in this area and hence of the agricultural and pastoral bases of its economic life. The area has been, and in many places continues to be, thinly populated in relation to land suitable for cultivation. Labour for the more arduous employments has therefore been hard to find, a feature which helps to explain the persistence into the twentieth century of slavery and other forms of enforcement of labour, and the population has been geographically mobile in response to changes in economic opportunities.

The simplest climatic contrast is between the forest and originally forested areas lying south of about 8°N and the more northerly savanna. The forest occurs close to the coast between Guinea-Bissau and eastern Nigeria, extending about 200 miles inland but interrupted for some 200 miles between Accra in Ghana and Porto Novo in Benin. It has a relatively heavy rainfall and long growing season, and tree-crops for export overseas have been extensively produced for several generations, especially in southern parts of Nigeria, Ghana and the Ivory Coast. Population density is related to these commercial possibilities and also to the conditions of food production, being higher in the areas of root-crop cultivation (yams, cassava, cocoyams) that lie east of Abidjan in the Ivory Coast than in the more westerly areas where rice is the staple food. The savanna is drier and has a shorter growing season. Its climate and the distance from ports of its interior have restricted, but not excluded, possibilities of agriculture for overseas markets. Its productivity in growing food crops (sorghum and millet, or mixtures of these cereals and roots) is markedly less than that of the forest.

A somewhat less crude division based on rainfall and the consequential length of the growing season for crop plants is of Sahelian, Sudanic, Guinean and other, intermediate, longitudinal belts.[3] The northernmost Sahelian belt has annual rainfall (unlike the Sahara) but the wet season is short, less than two-and-a-half months, and crops usually cannot be grown without irrigation; much of Mauritania and northern Senegal, Mali and Niger lie in this zone. In the Sudanic belt further south, the rainy season extends to about five months. In the sub-Sudanic belt, it is still longer, up to seven months, but this region has been less attractive to human settlement, partly because of the widespread occurrence of tsetse flies which makes cattle raising difficult or impossible. The sub-Guinean belt and the southernmost Guinean belt have two rainy seasons totalling more than seven months; the distinction between them is based on the high relative humidity of the coastal areas and on the nature of the regrowth of vegetation on fallow land, predominantly grasses in the sub-Guinean belt and shrubs and trees in the Guinean. A West Guinean subdivision, encompassing Guinea-Bissau, western Guinea, Sierra Leone and Liberia, experiences the heaviest rainfall and severest leaching of its soils. Of these climatic belts, the Guinean (including West Guinean) corresponds to the forest, the others to the savanna.

These climatic differences are powerful explanations of variations in economic opportunities and material levels of living within West Africa. Not only has food been more easily cropped in the forest, but comparative advantage could be exploited there in the production for overseas markets of palm oil and kernels, cocoa, coffee, rubber, bananas and timber. The incomes obtained in this way provided demand, and helped lay an infrastructure, for other commercialized activities – food-farming, urban services and manufacturing. Population was pulled toward the coast and large urban settlements grew up about the ports. In the drier and more remote savanna, the stimulus provided by overseas markets was less, though cash-cropping (of groundnuts principally) was extended rapidly in Senegal and northern Nigeria when cheap transport by rail became available, and some of the northern centres of population were so maintained and even grew. Poorest and least able to respond to changing opportunities have been the pastoralists of the Sahel and inhabitants of the sparsely populated lands intermediate between the Sudanic and Guinean zones, the so-called Middle Belt. Here are still to be found some of the materially poorest peoples in the world.

In recent times, the centre of gravity of West Africa has lain in the forest or on the coast, economically if not always politically. It was not always so. So long as West African traffic, both in manpower and products, lay mainly across the Sahara, it was in the Sudanic belt that large territorial states or empires were established and civilization was advanced. Reorientation of trade to the Atlantic shifted the balance toward the coast. It was the result initially of a new demand for slave labour in the Americas, and later of growing demand in industrialized Europe for tropical produce, especially oilseeds, and the fall in ocean freights in the late nineteenth century. African enterprise exploited these new opportunities; European colonialism when it came was an endorsement, rather than a direction, of the resulting transformation.

Mineral resources

The distribution of commercially exploitable mineral resources has been a secondary, but increasingly strong, determinant of variation in wealth in West Africa. Gold has long been quarried and panned in the region and was a principal export in the trans-Saharan trade; as late as the 1930s, many thousands of independent producers were engaged in its extraction. The mining and working of iron are also ancient in West Africa. But in recent times, and in contrast to agricultural production, mining has been undertaken mainly by large-scale 'expatriate' (non-African) enterprises; the chief exception has been in diamonds, where small-scale alluvial digging has persisted. The British colonial governments, with an eye as always to their revenues, found it expedient to confer rights of exploitation on expatriate companies in the important goldfields of the Gold Coast and in the tin mining area on the Jos Plateau in Nigeria. Later, large-scale mining of manganese and diamonds in the Gold Coast and of diamonds and iron ore in Sierra Leone was similarly instituted. The expatriate companies produced for export, had high ratios of overseas costs to total costs, and remitted abroad part of their profits and factor earnings. If it adds anything to call them enclaves, they were enclaves.

Minerals became important in the export trade of Sierra Leone and the Gold Coast but were not of much significance for West Africa as a whole until nearly the end of the colonial period, and the region did not appear to have great potential for mining. The pattern began to change after the Second World War. Developments have included large-scale mining of iron ore in Liberia and bauxite in Guinea since the 1950s; of iron in Mauritania and phosphates in Senegal and Togo since the 1960s; and of uranium in Niger since the 1970s. By far the most important has been mineral oil extraction in Nigeria, which began in 1957 and soon overshadowed the old agricultural and pastoral economy, accounting by the mid-1970s for about nine-tenths of the country's exports, three-quarters of its public recurrent revenues, and one-quarter or more of its GDP. Ghana became an oil producer in 1979, and the Ivory Coast in 1980; at least in the latter country, oil is expected to emerge as a major industry.

The links of the mining industries with other local activities, as purchasers or suppliers, have continued to be slight (except in oil refining), and their direct impact on the occupational distribution of labour has not been great. The chief nexus between their operations and the economic life of the countries in which they are located has been their payments to governments as royalties on production and taxes on or shares in profits. Depending as they do on the volume and profitability of production, these payments are variable and not always large. But in the case of Nigerian oil, where the gross value consists largely of economic rent, most of which has been publicly appropriated, and where both the volume and unit-value of production rose greatly after the 1960s, the receipts accruing to the state have been enormous relatively to the financial resources that would otherwise have been commanded.

Differences in levels of living: GNP estimates

Table 1 is an attempt to measure the differences in material levels of living among the countries of the West African mainland. It shows the estimated Gross National Product (GNP) of each country in 1979, both in the aggregate and averaged over the estimated population. GNP is a money valuation of all the exchangeable final goods produced and services rendered in a country during a year, less net payments made abroad for the use of foreign-owned factors of production. (The same valuation without allowance for earnings paid and received from abroad is the Gross Domestic Product or GDP.) The measurement is of only final goods and services because inclusion of intermediate products (such as the raw cotton used in manufacturing cloth or the fuel consumed in performing a transport service) would entail double-counting. Only exchangeable goods and services are taken into account on the ground that other things that people do for themselves or one another are non-economic in character, or not connected with the earning of livelihoods. The measurement is gross in the sense that no deductions have been made to provide for the maintenance intact of capital assets. (Where such depreciation allowances are made, the measurement becomes the Net National Product, sometimes called the National Income.)[4] The GNPs, which have been first estimated in national currencies, are shown in the table converted into a common denomination, the United States dollar, at the official rates of exchange ruling about 1979.

Table 1 Estimates of GNPs and national populations, 1979

	Population (million)	*GNP (US$ million)*	*GNP per head (US$)*
Ivory Coast	8.227	8,550	1,040
Nigeria	82.603	55,350	670
Liberia	1.797	900	500
Senegal	5.518	2,370	430
Ghana	11.313	4,530	400
Togo	2.420	850	350
Mauritania	1.588	510	320
Guinea	5.275	1,480	280
Niger	5.163	1,390	270
Benin	3.425	860	250
The Gambia	0.587	150	250
Sierra Leone	3.381	850	250
Upper Volta	5.642	1,020	180
Guinea-Bissau	0.779	130	170
Mali	6.750	940	140
West Africa	144.468	79,880	550

Source: World Bank, *Annual Report 1981*, p. 37.

The countries are listed in descending order of GNP per head, and it appears that for the Ivory Coast this quantity is more than seven times as great as for Mali. It is remarkable that, while ten of the fifteen countries fall within the World Bank's conception of 'low-income' countries in 1979 (GNP per head of $370 or less), the concentration of population in Nigeria, the Ivory Coast and Ghana produces for West Africa as a whole an average ($550) in what the Bank terms the 'middle-income' range. Those three countries account together for 71 per cent of the total West African population shown in the table, and for 86 per cent of the total of

GNPs. For Nigeria alone these proportions are 57 and 69 per cent respectively.

The West African average of $550 may be compared with estimates for the same year of $10,630 in the United States and $6,320 in the United Kingdom.

Reservations must be expressed concerning both the accuracy and the meaning of the comparisons made in Table 1 and of wider international comparisons of this sort. First, the statistics of production are unreliable. In each West African country, there are some activities for which fairly accurate information may be obtained because of the fewness of the enterprises engaged in production or marketing and their need to maintain detailed records for their own purposes. Examples are large-scale mining, construction and manufacturing, some public services like the railways and electricity supply, banking and possibly agricultural production for export. There are other activities where enterprises are usually small, their numbers very large and the channels of distribution various. They include farming and stock raising for home markets and for subsistence (i.e. consumption directly by the producers themselves), retail trade and many manufactures and personal services. Estimates of the volumes of these latter activities are more or less informed guesswork rather than based on collection of recorded information.

Further uncertainties arise when the estimated volumes of goods and services are valued. The average price of, say, yams in Ghana in 1979 can be computed with much less assurance than the average price of electricity. Other valuations are arbitrary. The value of public services like defence and schooling is what the government spends on them, since they are not purchased by their beneficiaries. The prices imputed to subsistence output – the output which is exchangeable in principle but not exchanged in practice – are bound to be fictive.

The West African GNP totals must therefore contain considerable margins of error. If the compilers of the national accounts have opinions on what those margins are, they usually do not confide them to the public, but some considerable revisions of estimates bear witness to the statistical uncertainties.[5] One authority has suggested an average margin of error in African GDP totals of 20 per cent (in either direction), in a range from 10 to 35 per cent, and a range of error in the sectoral components of the totals between 10 and 50 per cent (in either direction).[6]

The population totals are also unreliable. In a few West African countries, there has been no census of population; in others, only one census. A tendency for censuses to under-enumerate is sometimes corrected by adjusting the census total upward by as much as 15 per cent. In Nigeria, on the other hand, the censuses taken in 1963 and 1973 produced totals that were implausibly large in relation to the results of a previous census in 1952–3. While the 1973 census was in fact annulled (as another taken in 1962 had been), the dubious results of 1963 have provided the basis for all subsequent projections of the Nigerian population. Such is the uncertainty surrounding this matter that, while there is ground for supposing the figure of 82.6 million in 1979 to be an overestimate by as much as 15 million, some commentators have suggested it to be an underestimate. Since Nigerians are a substantial fraction of the population of West Africa by any reckoning, the failure to produce credible census results in Nigeria since independence makes the total for West Africa as a whole highly uncertain.

It does not follow from the unreliability of the population totals that the estimates of GNP per head are even weaker than those of GNP, for in practice the GNP is estimated partly on the basis of the population estimate. One should rather conclude that uncertainty about the true size of the population contributes to the uncertainty of the GNP estimate.

Second, international comparisons are affected by differences among countries in the uses of GNP. There are such differences in the proportion of the GNP which, being used in investment and in government consumption of goods and services for purposes such as defence, does not contribute immediately or directly to the levels of living enjoyed in households. Estimates of GNP per head may therefore mislead when used to compare individual material welfare, since they are rather measurements of average productiveness.

Third, even if deductions were made from the estimate of GNP per head to allow for investment and for government consumption of the kind mentioned, the reduced figure would remain only a quotient; it would not necessarily measure a modal or representative level of material living. Personal or household shares in the available product may vary considerably among occupational groups, between town and country, or among areas differentiated by climatic or other natural conditions. Intranational disparities may be as marked as the international disparities depicted in Table

1, and the table consequently abstracts from much of the variation in West African livelihoods.

Fourth, the international comparisons are vitiated by the practice of converting the national products, measured in national currencies, into a common denomination at the rates of exchange ruling with the US dollar. As has often been observed, the results of this practice can be astonishing. Thus, the $140 per head attributed as an average to Mali in 1979 is an annual income at which no inhabitant of the United States could survive, yet Malians not only survive but are said to be increasing their number at an annual rate of 2.6 per cent. Part of the explanation may be that the volumes of production in Mali are greatly underestimated. In addition, it would seem that the equivalent in Malian francs of $140 must have bought or represented more goods and services in Mali than $140 did in the United States. In other words, the exchange rate failed to reflect the relative purchasing power of the two currencies each in its own country. The main reason for this discrepancy would be the large volume of goods and services produced in each country that are not traded internationally and the relative values of which therefore cannot influence the exchange rate. Precisely because they cannot be sold in other, richer, markets, the prices of these so-called non-tradables are lower (relatively to the prices of the goods and services that are exported) in Mali than in the United States. Hence the Malian franc goes further in Mali than does its dollar equivalent in the USA.

It is argued on these grounds that GNP in poor countries is generally understated in relation to GNP in rich countries. It has been suggested that, if direct comparisons of real product were made, or if exchange rates were calculated on the basis of parity in the overall purchasing power of national currencies, increases in GNP in the poorest countries by a factor of about three would be in order.[7] Malian GNP 1979 might then appear as, say, $500 per head instead of $140.

If comparisons of GNP based on exchange rates are fallacious as between Mali and the USA, may they also be deceptive among the West African countries themselves? Because, again, not all the products of Mali and Nigeria are traded internationally, the equivalent in US dollars (or any other common denomination) of the Malian franc may be worth more in Mali than the equivalent in US dollars of the naira is worth in Nigeria. These disparities among West African countries in the overall purchasing power of their currencies are likely to be less than those between West African countries and much richer countries elsewhere. They may nevertheless be significant. Indeed, even within a country the purchasing power of the national currency may vary according to where it is spent and who spends it – a point that has relevance for estimates of income distribution.[8] The naira, for example, may go further in rural Borno than in Lagos. It may go further for the local man than for the stranger, and for the poorer than for the richer. The problem of comparability of monetary measurements of output or income becomes conspicuous when these measurements have to be converted into a common denomination, but it is present also when such conversions are unnecessary. Not only is the equivalent in Malian francs of a dollar worth more in Mali than the dollar is worth in America, but any currency has as many values as there are structures of relative prices facing its spenders.

The ranking of countries in Table 1 is probably not gravely misleading. It is plausible that the Ivory Coast, Nigeria and Liberia should appear at the top of the list. The first of these countries has been rapidly commercializing its agriculture on the basis of export markets since about 1950; the second has discovered wealth in oil; and the productiveness of the third is lifted by an iron mining industry large in relation to the economy of the country. The economic expansion of Senegal and Ghana has been slow in recent years, but they have long been reckoned among the most economically advanced parts of West Africa. It is unsurprising that Mali – far from the coast in the Sahelian belt, and without exploited mineral resources – should appear at the bottom of the list. But it may well be doubted, in view of the practical and conceptual problems of measurement that have been mentioned, whether the rough apprehensions long entertained of differences in the livings obtainable in the region are much refined by comparison of estimates of GNP per head.

Differences in levels of living: migration

These rough apprehensions depended partly on widespread evidence of migration in the region and of its connection with differences in natural conditions and economic opportunities. Movement of people, whether permanent or seasonal, in search of new land for cultivation or pasture or new grounds to fish, is a long-established feature of West African economic life. This mobility was enhanced in the colonial period by wider areas of common government, the gradual replacement of slave by free labour, and mechanical transport. Large numbers of men moved over long distances, often temporarily. They came especially from areas north of about the tenth parallel – from the valley of the river Senegal, Portuguese and French Guinea, Soudan, Upper Volta, Niger, northwestern and northeastern Nigeria and the Northern Territories of the Gold Coast. Togo was also an important source of migrants. Their principal destination was employment as share-croppers or wage-labourers in commercialized agriculture, especially in the forest zones of the Gold Coast, the Ivory Coast and southern Nigeria and in the groundnut-growing areas of Senegal, The Gambia and northern Nigeria. The periodicity of movement was largely governed by the long dry season in the northern savannas. Employment was also found, though initially it was less preferred, in public works, urban services and mining. Numerous men and women working on their own account as traders also moved over long distances.

Influx and efflux of labour on a large scale were already taking place in Senegal and The Gambia as early as the middle nineteenth century. By the 1950s, this movement was reckoned among the most conspicuous features of the West African economic scene. Estimates were then current of 300–400,000 persons moving annually between the Gold Coast and the neighbouring French territories, 200,000 between the Northern Territories and other parts of the Gold Coast, 60,000 into and out of the cocoa-growing areas of western Nigeria, and 40–55,000 between Senegal and The Gambia and the surrounding territories. There is an estimate of 140,000 migrants entering the Ivory Coast in 1959. A census of outward migration taken in Sokoto Province in northwestern Nigeria during the dry season of 1952/53 recorded the passage of 259,000 migrants, whose destinations were mostly elsewhere in Nigeria but also included the Gold Coast.[9] The size of these movements has probably diminished since the end of the colonial period, especially in Ghana, but a more recent estimation still arrives at an annual movement between the interior and the coast of about 300,000 persons, of whom two-thirds might be seasonal and one-third permanent migrants.[10]

In addition to these estimates of annual movements, the population censuses have provided measurements of the foreign and migrant elements in national populations. About 1975, there were

2.8 million foreign nationals (7 per cent) in an aggregate population of 40 million in nine of the more westerly countries – Senegal, The Gambia, Mali, Sierra Leone, Liberia, the Ivory Coast, Upper Volta, Ghana and Togo – and another 4.4 million (11 per cent) were classed as internal migrants; these proportions would be higher for the economically active population, or the labour force – about 11 and 18 per cent respectively.[11] Nigeria is excluded from the measurements for want of usable census data, but there was undoubtedly much movement of population in that country in the 1970s, both internal migration and immigration from Benin, Togo, Ghana, Niger, Chad and Cameroon.

In the Ivory Coast, since 1960 the largest recipient of immigrants, the census of 1975 recorded 1.43 million foreign nationals (21 per cent of the total population of 6.77 million), of whom 1.054 million were foreign-born. About half of these foreigners were nationals of Upper Volta, and there were some 350,000 Malians, 100,000 Guineans, 50,000 Nigerians and 40,000 Ghanaians. The census also enumerated 1.265 million Ivorian nationals (19 per cent of the population) living outside the administrative departments in which they had been born. The principal region of out-migration is the North Region, contiguous to Upper Volta and Mali. Among men aged between fifteen and sixty-four years, 30 per cent were foreign nationals, 24 per cent were inter-departmental internal migrants, and only 35 per cent were living in the localities in which they had been born.

In Ghana, there were 562,000 foreign nationals (7 per cent) in a 1970 census population of 8.559 million, and 350,000 of them were foreign-born. They included 247,000 Togolese, 157,000 nationals of Upper Volta, and 56,000 Nigerians. There had been a larger foreign element in the population ten years earlier – 12 per cent according to the 1960 census, possibly 15 per cent at the peak of seasonal immigration in that year[12] – but a wholesale expulsion of aliens, affecting particularly the Nigerians, occurred at the end of 1969. Of the 1970 census population, 1.4 million (16 per cent) were Ghanaian nationals enumerated in regions other than those in which they had been born. The principal regions of out-migration were the Upper and Northern Regions in the savanna south of Upper Volta and the Volta Region adjacent to Togo.

Such demographic data as are available for the analysis of West African migration indicate a positive relationship between the rate of immigration and the economic status of a country represented by its GNP estimate. A positive relationship is also observable between the rate of in-migration and economic conditions among the administrative divisions of individual countries.[13]

West Africa's international trade

Movements of labour – both permanent and temporary, within and across territorial or national boundaries – have been an indispensable condition of economic growth in West Africa and a means by which the benefits of that growth have been diffused. Redistribution of labour to the naturally more favoured areas was demanded by the labour-intensive character of agricultural production for export overseas and of the trade and transport, building and urban services which export agriculture fostered. Even mining (other than oil extraction) would not usually have been possible on the scale achieved but for the readiness of labour to migrate.

Considered in relation to total economic activity, exports do not appear of overwhelming importance in the West African region as a whole, or in most of the countries individually (the exceptions are Liberia and The Gambia). In 1979, the value of exports of the fifteen mainland countries was recorded as about US $24.3 billion, or 23 per cent of the total of their GDPs. This ratio is not remarkably high, having regard for the small size of West Africa relatively to world production and to the comparatively undiversified economic life of the region. Thus in the UK, the dollar value of whose GDP in 1979 was nearly four times the combined total for West Africa, the ratio of exports to GDP was also 23 per cent.

Even so, in West Africa, as in many other regions, exports have served historically as prime mover in economic growth, since the earnings from resources were commonly higher when they were used to supply distant markets than when their utilization met only local demand. Both for the region as a whole and in individual countries, exporting capacity and economic growth have been strongly connected.

These exports have been and still are largely to markets overseas. Of the recorded exports of West African countries in 1979, less than 3 per cent represented their exports to one another. Admittedly, international trade within the region is under-recorded. There is much illicit trade across frontiers to evade customs duties and official controls on prices and on access to foreign exchange. But this smuggling is largely transit trade within West Africa of goods destined for or obtained from countries overseas. Its absence from the record is therefore scarcely germane to the present discussion. Also insignificant in the present context is an under-recording of the border trade that consists in selling local foodstuffs in markets physically convenient to sellers but technically on different national territory from their own; this trade is only an incidental extension of local commerce across frontiers. There is, finally, more genuine international trade in local products in the region, taking advantage of differences in resources or natural conditions. This trade includes the export of crude oil from Nigeria to refineries elsewhere on the coast, and of petroleum products from the Ivory Coast to interior countries. Another part is the north–south trade including livestock on the hoof, dried fish, kola nuts, vegetables and yams. Even if the true value of this latter trade were, as is sometimes suggested, three times as great as the value recorded, it would do little to remove the impression of a predominantly overseas orientation in the exports of West African countries.

International trade within West Africa is relatively little developed because the countries of the region are not economically complementary in any marked degree; their structures of demand and production are not such that they can provide important markets and sources of supply for one another. Nor has such complementarity been pronounced within West African countries, as among the districts of each. Consequently, most internal commerce in home-produced goods is local rather than national in scope. The growth of large-scale manufacturing in West Africa since the late 1950s has diminished the local character of trade in local products within countries of the area, and it has encouraged plans similarly to widen markets in West Africa as a whole. But as yet exporting capacity, even in manufactures, continues to be more readily absorbed overseas than in neighbouring countries.

Two kinds of economies

If an economy is understood to be a cosmos, or spontaneous order maintained without conscious volition, in which the getting of livelihoods is indirectly coordinated by impersonal incentives and deterrents, West Africans like other peoples live in several overlapping economies. For many of the majority who live in rural areas, the most important of these economies is local, in the sense that they sell their products or services to meet local demand and obtain most of their requirements from local sources. But participation in wider economies is almost universal in the region, and of greater importance than local participation for many urban residents. These wider economies extend as far as the entire world, or such of it as trades with West Africa. Overall, and as compared with richer and more economically diversified regions, the narrowest (local and district) and widest (world) economies appear still to be of greater importance in West Africa relatively to the intermediate (national and regional) economies – though this contrast is lessening over time. National and regional markets developed sooner in West Africa for factors of production (notably labour) than for goods and services.

Alternatively, an economy may be defined as an organization deliberately constructed to serve a consciously formulated objective or hierarchy of ends. Here again we encounter a multiplicity of units of analysis. Indeed, there are as many such economies as there are economic factors. Economies thus understood as businesses or enterprises can be identified at every level from individuals and households to national governments and transnational corporations. Among them, national governments are of particular importance and interest because their objectives extend to the management of economic life within the boundaries of the territories they administer, thus producing a physical correspondence, but not necessarily harmony, between the economy of a country considered as a cosmos (or part of a wider cosmos) and the 'national economy' understood as a collective enterprise directed by a government. The later chapters of this book are concerned with the attempts of governments to manage economic life in West Africa, and in particular with their plans of development. But first the economic anatomy of the region is further pursued. The next chapter explores in more detail the economic structures of West Africa, Chapter 3 its human resources and Chapter 4 its external trade.

Summary

This book is concerned with West Africa, considered as the area lying west of Cameroon and south of the Sahara – a delimitation owing more to colonial administrative history and diplomatic alignments than to geography or ethnography. Economic opportunities and levels of living in this area depend heavily on climatic and other natural conditions, and especially on the amount and incidence of rainfall. These conditions are most adverse in the Sahelian belt in the north and in areas intermediate between the Sudanic belt and the forested areas of the Guinea coast. The distribution of commercially exploitable mineral resources has been a secondary, but increasingly powerful, determinant of economic disparities among West African states.

These disparities can be represented by estimates of GNP per head, but the measurements both of production and of population are shaky, the averages per head do not denote modal levels of living, and there are technical difficulties in comparing monetary magnitudes among (and even within) countries. That there have been, and continue to be, substantial disparities is nevertheless evidenced by large-scale migration of labour both within and across territorial or national boundaries.

These movements of labour have made a crucial contribution to export trade, which has been the prime mover of economic growth in West Africa as in many other parts of the world. The exports are very largely to markets overseas. International trade within West Africa is little developed because there is little economic complementarity among the countries of the area. Even their internal trade tends to be localized, though the growth of large-scale manufacturing is diminishing this tendency.

West Africans live in many overlapping economies, whether we define economy as a spontaneous order in which economic activities are indirectly and impersonally coordinated, or as an enterprise pursuing consciously formulated objectives. The assumption that economies (in either sense) are conterminous with national boundaries is appealing to governments that seek to manage and develop economic life, but greatly oversimplifies reality.

2

Economic Structures

National accounting

The structures of national economies are described by the national accounts compiled in estimating GDPs. These accounts show the composition of a country's economic output, the uses made of this output, and its distribution among factors of production. In West Africa, national accounting has been practised since the early 1950s, and for Nigeria and Ghana there are now long (but not necessarily consistent) series of annual estimates. It must be said at once that the information given by these accounts is by no means exact and that the purposes to which it can reasonably be put are severely limited by its unreliability.

The GDP has already been defined as a money valuation of all the exchangeable final goods produced and services rendered in a country during a year. The most obvious way of arriving at this aggregate is by measuring the net output or 'value added' by each producer (i.e. the value of his output less the cost of other producers' output which he uses in producing his own). These net outputs can be valued either at the prices buyers pay for them (or supposedly would pay for them if they were offered for sale), known as market prices, or at the prices producers receive for them (or supposedly would receive for them if they were offered for sale), known as factor cost; the two prices diverge if a tax is being levied or a subsidy paid on the product concerned.

The sum of the net outputs is thus the GDP measured either at market prices or at factor cost, and the former measurement exceeds the latter by the amount of indirect taxation minus subsidies. The division of the total among economic activities can then be shown in less or more detail, distinguishing broad sectors such as agriculture, mining, manufacturing and services, or narrower sections such as particular crops and manufacturing industries. Since the initial impact of indirect taxes is mostly on distributors, the relative importance of distributive trade (and of the wider sector of services) is likely to be magnified by a measurement at market prices as compared with factor cost.

The GDP may also be measured from statistics of expenditure. The purchases (in practice or in principle) of every kind of final product are summed and the total adjusted for changes during the year in the stocks (or inventories) of goods produced but not yet bought. This method gives a GDP at market prices, which can be reduced to factor cost by subtracting the amount of net indirect taxation. A summary statement of the GDP measured in this way distinguishes among private consumption, government consumption, gross fixed capital formation, changes in stocks, and exports, but the sum of these items exceeds the GDP by the amount of spending on imports which has therefore to be deducted; for want of information this deduction is usually made from the total of expenditure instead of being allocated among the items distinguished. Capital formation denotes spending on tangible assets, like buildings and machines, whose potential usefulness is exhausted over a period of years; other domestic spending, including all the expenditures of households other than on housing, is classed as consumption. The distinction between these categories has a theoretical basis but its observance in national accounting is largely conventional.

A third method of computing the GDP is by summing the incomes obtained (in cash and kind) from producing goods and services. This approach yields totals of wages and salaries earned, rents received, the gross profits of corporate businesses (private and public), and earnings from self-employment. These are factor incomes measured before tax and excluding receipts such as pensions not earned from current production. Their sum is the GDP at factor cost.

A summary tabulation of a complete set of GDP estimates might therefore take the form shown in Table 2, where the sectoral origin is obtained by the output method of computation, the uses of GDP by the expenditure method and the factor shares by the income method. Economic statistics so collected and organized will apparently show both the relative importance in any year of various branches of economic activity, uses of output and factor shares in income, and the changes in these proportions between one year and another.

In practice, in no West African country are economic statistics

Table 2 Categories of GDP

Sectoral origin of GDP	*Uses of GDP*	*Income shares in GDP*
Agriculture	Private consumption	Wages, salaries and supplements
Mining	Govt consumption	Rents
Manufacturing	Gross fixed capital formation	Gross trading profits of private companies
Electricity, gas and water	Changes in stocks	Gross trading profits of public corporations
Construction	Exports of goods and services	Earnings from self-employment
Transport and communications	Total final expenditure	
Public administration	*Less* imports of goods and services	
Distribution	GDP at market prices	
Other services	*Less* net indirect taxes	
GDP at factor cost	GDP at factor cost	GDP at factor cost

adequate to allow estimation of the GDP independently by each of the three possible methods, or presentation of the national accounts fully in accordance with the categories used in Table 2. As was observed in Chapter 1, the accuracy of such results as are obtained is unlikely to be high. In the output account, information is relatively firm for large-scale mining, manufacturing, transport and construction enterprises and for public utilities and administration – though relative firmness here could mean margins of error of 10, or even 20, per cent.[1] Information is much more unreliable (margins of error up to 50 per cent) for agriculture, the trade in agricultural products in home markets, and small-scale manufacturing and services, which together account for at least one-third and in some countries for over one-half of the GDP estimate. The information is so weak for these sectors because the producers concerned are largely unenumerated and do not maintain records of their activities. The output attributed to them includes the imputed value of production which is exchangeable in principle but has not been exchanged in practice; this 'subsistence output', whose valuation must be arbitrary,[2] accounts for between 10 and 25 per cent of most estimates of GDP in West Africa.

In the expenditure account, the figures for government, both fixed capital formation and the other expenditures labelled as consumption, might be assumed relatively good, although the ability of governments to keep record of their own transactions is much weaker than might be supposed. Fixed investment by large-scale enterprises can also be ascertained, but investment elsewhere is a matter of conjecture from statistics of the production and importation of cement and other building materials and producers' goods. Non-monetary investment, or the subsistence output that consists in clearing land and in building with traditional materials, is difficult to estimate and may not be taken into account. Figures of lawful export sales and import purchases are obtainable, but smuggling (including false invoicing) is sometimes relatively large,[3] and the valuation of it, so far as it is valued, can be only approximate. Information on changes in stocks may be available only for export commodities. (In Nigeria, this item is not distinguished from the remaining category of private consumption.) Some private consumption expenditures, such as on cigarettes and petroleum products, might be accurately measured because of the fewness of the producing units or the distributive channels, but private consumption in the aggregate is likely to be a residual, the result of subtracting the estimates of government consumption, capital formation and exports less imports from a GDP already computed. This residual is nearly always the greater part of the GDP and sometimes as much as 80 per cent.

Finally, in the absence from West African countries of detailed information on personal incomes such as might be provided by a comprehensive system of direct taxation, estimation of GDP by the income method encounters the same difficulties as the output method. The gross profits of the larger enterprises and their wage and salary payments (along with those of the government) can be ascertained, but information is as uncertain about other incomes as it is about the net output of small-scale enterprises in agriculture, manufacturing and services. As compared with sectoral origin, therefore, income shares are an alternative way of stating the composition of the GDP but scarcely an alternative way of computing the total.

These shortcomings in the quality of aggregated economic

Table 3 Sectoral origin of GDPs in ten West African countries, about 1978 (*percentages*)

	Benin 1978	*Ghana 1977*	*Ivory Coast 1978*	*Liberia 1977*	*Mauritania 1978*	*Nigeria 1977/78*	*Senegal 1979*	*Sierra Leone 1978/79*	*Togo 1979*	*Upper Volta 1974*
Agriculture	42.5	56.2	24.5	13.6	25.0	23.1	25.1	34.2	27.6	41.4
Mining	0.2	0.8	0.2	20.4	12.1	24.4		11.8	8.3	0.1
Manufacturing	5.4	10.8	11.8	7.1		4.8		7.6	5.3	10.3
Electricity, gas and water	0.9	0.5	1.2	1.0	5.7	0.3	19.7	0.5	1.6	1.6
Construction	3.3	3.8	8.8	6.2	3.9	9.2	4.4	3.0	7.9	4.8
Total: industry	9.8	15.9	22.0	34.7	21.7	38.7	24.1	22.9	23.1	16.8
Transport and communications	7.5	3.0	6.9	7.6	8.4	3.2	6.2	11.3	6.1	7.6
Other services	40.3	25.0	46.5	44.0	44.9	34.9	44.7	31.6	43.0	34.4
Total: services	47.8	28.0	53.4	51.6	53.3	38.1	50.9	42.9	49.1	42.0

Source: Calculated from data in UN, *Monthly Bulletin of Statistics*, October 1981, Special Table F.

statistics in West Africa do not make the computation of GDP useless, but they argue for caution in the purposes to which the estimates are put and in the interpretations placed on them.[4] It would not be surprising if there were a common tendency to underestimate the volume and values of these non-formalized activities for which the evidence is fragmentary or indirect, and hence to exaggerate the relative importance of formalized and enumerated activities. Comparisons of ratios between countries, or over a lengthy period of time in the same country, may mislead both because of variations in the quality of information among places and times and because of differences in the definitions adopted and the assumptions made by the responsible statisticians. Small differences in a ratio between one year and another may have no real significance. As will appear later in this chapter, margins of error in the estimates become particularly damaging if rates of growth in the GDP and its principal sub-aggregates are calculated.

Sectoral origin of GDPs

Table 3 shows estimates published by the Statistical Office of the United Nations of the sectoral origin of market-price GDP in ten of the West African countries about 1978. A division is made among primary, secondary and tertiary activities. The first comprises agriculture including livestock, hunting, forestry and fishing; the second embraces mining, manufacturing, construction and the public utilities; and the third includes transport and communications, distribution, public administration, and other services.

The figures in the table are perhaps as much revealing of differences among the countries in methods of computing GDP, including the delimitation of sectors and the valuation of subsistence output, as of real contrasts in economic structures. Thus, it is difficult to believe that the relative importance of primary activities is twice as great in Ghana as in the Ivory Coast, Mauritania and Togo. It strains credulity that the share of transport and communications can be over three times as great in Sierra Leone as in Ghana and Nigeria. The implication of the percentages shown against manufacturing, that industrialization has proceeded as far in Upper Volta as in Ghana and the Ivory Coast, and much further than in Nigeria, is surely questionable. The valuation of GDP and its sectoral allocation would seem to depend heavily on national statistical conventions, and many of the differences among countries shown by the table would seem to be spurious.

Some of the contrasts do have validity. The abnormally large shares of construction in GDP in the Ivory Coast, Liberia, Nigeria and Togo reflect high rates of fixed capital formation in those countries. The differences in the relative importance of mining also have substance. One-quarter of the Nigerian GDP and one-fifth of the Liberian are shown in the table as attributable to mining. The contribution of this sector was also substantial about 1978 in Mauritania, Sierra Leone and Togo (as in Guinea and Niger, which are not shown in the table), while it was small in Benin, Ghana, the Ivory Coast and Upper Volta (and in The Gambia, Guinea-Bissau and Mali). In Nigeria and Liberia, the emergence of large mining industries mainly explains the low proportions of GDP attributed to agriculture, the growth of agricultural output having fallen far behind that of the minerals.

A genuine contrast may also be made between all the West African countries and economically advanced countries. Thus, while primary activities account for between one-quarter and one-half of most West African estimates of GDP, in the UK in 1978 they contributed only 2.3 per cent, and in the USA 2.9 per cent. Secondary activities constituted 35.8 per cent of the British GDP estimate and 34.0 per cent of the American, and this was mainly because of high manufacturing proportions (24.8 and 24.2 per cent respectively).

Relative contraction in agriculture and increase in manufacturing are to be expected in the course of economic growth, since income-elasticities of demand are generally higher for manufactures than for foodstuffs. The possibilities afforded by international trade may retard the growth of manufacturing in a country in response to rising incomes, but they are unlikely to prevent it indefinitely. Changes in the relative importance of tertiary activities in the course of economic growth are less predictable. It is probable that the greater part of output is obtained as services in both the poorest and the richest societies, and that there is an intermediate range in which manufacturing grows relatively to tertiary as well as to primary activities.

In some West African countries, striking changes in the composition of output can be observed in the period for which national accounts are available, but they do not appear to have resulted chiefly from increase in manufacturing.

In Nigeria, primary activities represented 66 per cent of the GDP

Table 4 Structures of production in West African countries, 1960 and 1979 (percentages of GDP estimates)

	Benin		*The Gambia*		*Ghana*		*Guinea*		*Guinea-Bissau*		*Ivory Coast*		*Liberia*	
	1960	*1979*	*1960*	*1979*	*1960*	*1979*	*1960*	*1979*	*1960*	*1979*	*1960*	*1979*	*1960*	*19*
1 Agriculture	55	43	43	46	41	66	56	41	n/a	54	43	26	40	3.
2 Industry	8	12	18	9	19	21	36	26	n/a	9	14	23	37	2
2a Manufacturing	*3*	*8*	*n/a*	*n/a*	*10*	*n/a*	*n/a*	*5*	*n/a*	*n/a*	*7*	*12*	*n/a*	[illegible]
3 Services	37	45	40	46	40	13	8	33	n/a	34	43	51	23	3
4 2a as a percentage of 1+2	5	15	n/a	n/a	17	n/a	n/a	7	n/a	n/a	12	25	n/a	1

in 1950/51, the first year for which an estimate is available,[5] or 73 per cent if the profits of the statutory marketing monopolies of export crops are credited to agriculture instead of to distribution. By 1960/61, this proportion (including the marketing board profits) had fallen to 63 per cent, by 1970/71 to 49 per cent and by 1977/78 (according to Table 3) to 23 per cent. The main, though not the only, cause was the expansion of mining from about 1 per cent of the GDP estimates in the 1950s to 10 per cent in 1970/71, after the emergence of the oil industry, and 24 per cent in 1977/78.

A wider view of these changes over time can be obtained from Table 4, which reproduces World Bank estimates of the breakdown of West African GDPs among primary, secondary and tertiary activities in 1960 and 1979. In addition, the ratio of manufacturing to GDP is shown separately and, to obtain what is sometimes regarded as a better index of industrialization,[6] manufacturing is also shown as a ratio of total commodity production, or of GDP less services.

An indication of the uncertainties afflicting these estimates can be obtained by comparing the 1979 figures in Table 4 with those for the same or adjacent years derived from another source and shown in Table 3. Quite substantial discrepancies can be observed in the

Mali		Mauritania		Niger		Nigeria		Senegal		Sierra Leone		Togo		Upper Volta	
960	1979	1960	1979	1960	1979	1960	1979	1960	1979	1960	1979	1960	1979	1960	1979
55	42	59	27	69	44	63	22	24	29	n/a	36	55	25	62	38
10	11	24	33	9	32	11	45	17	24	n/a	23	16	23	14	20
5	*6*	*3*	*8*	*4*	*10*	*5*	*5*	*12*	*19*	*n/a*	*5*	*8*	*7*	*8*	*14*
35	47	17	40	22	24	26	33	59	47	n/a	41	29	52	24	42
8	11	4	13	5	13	7	7	29	36	n/a	8	11	15	11	24

otes: Agriculture includes livestock, hunting, forestry and fishing; industry comprises mining, anufacturing, construction and electricity, water and gas; all other economic activities are classed as rvices.

a indicates not available.

urces: World Bank, *World Development Report 1981*, Annex Table 3, supplemented by World ank, *World Development Report 1980*, Annex Table 3, and World Bank, *Accelerated Development in b-Saharan Africa* (1981), Annex Table 3.

breakdown of Senegalese and Togolese GDPs, where the two sets of estimates refer to the same year. Even more puzzling are the divergences in Liberia and Ghana between the 1977 (United Nations) figures given in Table 3 and the 1979 (World Bank) figures given in Table 4. In Liberia, seemingly, the respective shares in GDP of agriculture, industry and services changed from 14, 35 and 52 per cent in 1977 to 35, 26 and 39 per cent in 1979; and in Ghana from 56, 16 and 28 per cent in 1977 to 66, 21 and 13 per cent in 1979. In the Ghanaian case, the mystery is deepened by the World Bank's own, previously published, estimates for 1978, according to which the shares were 38, 18 and 44 per cent respectively.[7]

Whatever the validity of the figures given in Table 4 may be, they show for eight countries marked contractions between 1960 and 1979 in the relative importance of agriculture, balanced by expansion in industry or services or both. Exceptional cases appear to be The Gambia, where the relatively contracting sector has been industry (though the figure given for 1960 is improbably high); Ghana and Senegal, where it has been services, suggesting a declining ratio of commercialized output; and Guinea and Liberia, where industry as well as agriculture has contracted in relation to GDP.

On a weighted average in 1960 (the weights being the World Bank's estimates of the national GDPs in dollar terms), agriculture contributed 53 per cent of West African GDP, industry 15 per cent (including manufacturing, 6 per cent) and services 32 per cent. In 1979, using the GDP estimates of that year as weights, the averages were 28 per cent for agriculture, 38 per cent for industry (including 6.5 per cent for manufacturing) and 34 per cent for services. Overall, therefore, there has been a relative contraction in agriculture balanced by relative expansion in industry – but not so much manufacturing industry as mining and construction. To a large extent, these changes reflect the changing structure of the Nigerian GDP estimate and the growing importance of Nigeria in the combined West African GDP (from a share of about 45 per cent in 1960 to 70 per cent in 1979).

The weighted average ratio of manufacturing to GDP in 1979 was 6.5 per cent, but apparently in Senegal the importance of manufacturing was nearly three times as great. Considerable increase in this manufacturing ratio since 1960 is estimated for several countries, but it has been less important as an explanation of the relative decline of agriculture than the growth of mining and than an increased commercialization of economic life (including agriculture) that has raised the share of distribution, transport and other services in the GDP of nearly all the countries.

The weighted average ratio of manufacturing to total commodity production (or GDP less services) is, of course, higher, but it rose only modestly between 1960 and 1979 – from about 9 to 10 per cent. If minimum values of 20 and 40 per cent in this ratio are used to distinguish 'industrializing' and 'semi-industrialized' from 'non-industrial' countries,[8] it will be observed from Table 4 that Senegal by 1979 was not far short of semi-industrialized and that Upper Volta as well as the Ivory Coast was industrializing. Nigeria, on the other hand, remained emphatically non-industrial – much more so than Benin, Togo, or even Mali. It is unlikely that this classification of West African countries corresponds with reality, or in other words that the statistics showing the structures of national production are really comparable; in particular, the tendency of francophone countries to appear more industrialized than the anglophone must be suspect. (It is instructive that Ghana, the Ivory Coast, Nigeria and Senegal are recognized under the trade liberalization programme of ECOWAS as industrially more advanced countries, and the other member-countries as industrially less advanced.)

The apparent retrogression of the Ghanaian economic structure is remarkable in view of strenuous governmental efforts to develop it, especially in the 1960s. Apparently, secondary activities have remained roughly constant at about one-fifth of GDP in a twenty-year period, while primary activities, far from contracting, have expanded in importance. In contrast, the estimates indicate substantial structural change of a more positive kind in the Ivory Coast, where official policies have been less clearly directed at this result. Yet it would be erroneous to conclude that the Ivory Coast had ceased (while Ghana continued) to be predominantly an agricultural country. Trade margins, transport charges and taxes on agricultural products make up a considerable part of the output attributed to the tertiary sector, and over half the value added in manufacturing in 1975 resulted from processing agricultural raw materials.[9]

Inter-sectoral connections

The sectors of West African economies are weakly connected. This feature does not appear in the analysis of GDP, where only final goods and services are counted, but emerges when the destinations of all output, intermediate as well as final, are estimated. Thus, a simple input–output table prepared for Ghana in 1960 and distinguishing ten sectors showed inter-sectoral sales to be only about 8 per cent of the estimated value of all sales.[10] The remaining 92 per cent was sold directly to final domestic users for consumption or investment, added to stocks, or exported. In Nigeria in the same year, distinguishing twenty sectors, the ratio of inter-sectoral sales was about 12.5 per cent.[11] In 1968, an eight-sector table for Ghana produced a ratio of 13.2 per cent, and in 1977 the ratio of inter-sectoral sales in that country was officially stated to be little more than 20 per cent, but on an unstated basis of measurement.[12] The Nigerian ratio has presumably also risen because of increasing local industrial use of crude oil, natural gas, rubber, timber and cotton. But undoubtedly these ratios would still be low in West Africa by comparison with more economically advanced countries, and they are therefore sometimes said to be revealing of the extent of underdevelopment.

The ratios depend to some extent on the chosen sectoral division of economic activity. A finer division (separating, say, the processing from the growing of local foodstuffs, or the manufacture of clothing from textiles) would produce a higher ratio of intermediate to total sales. But the main determinants of the ratio are the kinds of output produced and whether they are exported or used locally. On the one hand, much production in West Africa exerts little demand on locally produced inputs, either because of the simplicity of the technology (as in agriculture) or because the inputs it needs are available only by importation (as frequently in large-scale mining and manufacturing). On the other hand, much of the production of goods that require further processing before they are suitable for final use (like crude oil and cocoa beans) is destined for export, and the processing can be more economically performed outside West Africa. The low ratios of intermediate output therefore reflect the current comparative advantage of West African countries in international trade. They also reflect a feature suggested in Chapter 1, the high affiliation of West Africans with local economies and the world economy relatively to national economies.

Uses of GDPs

Turning from the sectoral origin of GDP to its uses, Table 5 shows ratios in ten countries about 1978 for the five categories of expenditure previously distinguished (private consumption, government consumption, gross fixed capital formation, increase in stocks, and exports). The expenditures underlying each ratio are inclusive of spending on imports, the total of which then appears as a negative entry to reduce total expenditure on final goods and services to equality with the GDP at market prices. The five ratios themselves can therefore aggregate less or more than 100, according as to whether foreign trade makes a net subtraction from or addition to the goods and services available domestically (or whether the so-called 'resource balance' is positive or negative). Where exports (domestic output not retained at home) exceed imports (foreign output added to retained domestic output), the difference is a positive resource balance used to buy external assets or to reduce external liabilities. Where imports exceed exports, as in all but one of the instances shown in Table 5, the difference is a negative resource balance representing foreign production which has been received as loans, grants or direct foreign investment or through the sale of external assets.

The figures given in the table suggest great variation among West

Table 5 Uses of GDP in ten West African countries, about 1978 (*percentages*)

	Benin 1978	*Ghana 1977*	*Ivory Coast 1978*	*Liberia 1980*	*Mauritania 1977*	*Nigeria 1978/79*	*Senegal 1978*	*Sierra Leone 1978/79*	*Togo 1979*	*Upper Volta 1978*
Private consumption	98.8	77.4	53.8	55.2	54.6	63.6*	79.1	87.0	65.0	89.5
Government consumption	8.7	12.6	16.0	15.8	35.7	12.6	16.6	6.4	15.2	12.6
Gross fixed capital formation	14.0	9.4	30.5	25.5	40.2	31.4	13.7	14.3	47.6	21.8
Increase in stocks	3.3	1.7	0.4	3.6	5.8		3.8	0.9	2.4	3.6
Exports	23.9	10.5	37.0	53.4	33.6	25.7	37.6	23.8	25.2	14.6
Imports	−48.7	−11.5	−37.7	−53.4	−69.9	−33.4	−50.7	−32.4	−55.4	−42.1
Resource balance	(−24.8)	(−1.0)	(−0.7)	(–)	(−36.3)	(−7.7)	(−13.1)	(−8.6)	(−30.2)	(−27.5)
GDP at current market prices	100.0	100.0	100.0	100.0	100.0	100.0	100.0	100.0	100.0	100.0

Note: * Including increase in stocks

Sources: Calculated from data in IMF, *International Financial Statistics Yearbook 1981*, and UN, *Monthly Bulletin of Statistics*, October 1981, Special Table D.

African countries in the three main ratios of domestic expenditure. Private consumption is shown ranging between little more than one-half of the GDP in the Ivory Coast, Liberia and Mauritania and nearly 100 per cent of the GDP in Benin. The share of government consumption appears two-and-a-half-times as great in the Ivory Coast, Liberia and Senegal as in Sierra Leone, and over five times as great in Mauritania. The investment rate is less than 15 per cent of GDP in Benin, Ghana, Senegal and Sierra Leone, and more than 30 per cent in the Ivory Coast, Mauritania, Nigeria and Togo. The negative resource balance, or import surplus, is small in Ghana, the Ivory Coast and Liberia, but one-quarter or more of the GDP in Benin, Mauritania, Togo and Upper Volta.

Considerable year-to-year variations in these ratios can occur as a result of fluctuations in the resource balance. A sudden rise in export earnings, for example, may be associated with a sharp contraction in the ratio (though not the absolute amount) of private consumption if, as happened in Nigeria in 1974, a large export surplus accrues.[13] An import surplus financed by large-scale foreign direct investment may produce a jump in the ratio of fixed capital formation, giving to a country with a small GDP an investment rate that appears incongruous with its estimated income per head.

There has nevertheless been a trend since the Second World War for fixed investment and government consumption to rise relatively to private consumption – but not a trend equally pronounced or sustained in all countries. One reason for a rising investment rate is that West African governments since the war have been recipients of foreign aid, mostly tied to fixed capital formation; some of them have also, at times, borrowed heavily abroad on commercial terms to finance investment in the public sector. In addition, some of the countries have become more attractive to direct foreign investment in mining and large-scale manufacturing. Government spending on both capital formation and current goods and services has grown relatively fast since development was accepted as an official responsibility in the 1940s. This growth was accented by the processes of decolonization and the accession of West African countries to political independence, and it has been partly sustained in the better-off countries (Nigeria and the Ivory Coast) by efforts to distribute public amenities and services more widely among the populations. In contrast, private consumption has lacked powerful propelling forces; a factor that could at times have caused its rapid increase, the growth of agricultural export earnings, has been controlled by taxation imposed principally through official marketing monopolies.

Nigeria is the clearest illustration of the trend, since all series of GDP estimates for the country, at both current and constant prices, show government consumption and fixed investment rising much faster than private consumption.[14] According to the series reproduced by the IMF, government consumption rose from 3 per cent of the GDP in 1950/51 to 6 per cent in 1960/61, 10 per cent in 1970/71 and 13 per cent in 1978/79. For the same years, fixed capital formation appears as 6, 11, 16 and 31 per cent respectively. Over the whole period, the estimates show private consumption falling from 87 to 64 per cent.[15]

Ghana presents a different pattern. Government consumption rose fairly steadily as a proportion of GDP from 6 per cent in 1950 to 14 per cent in 1965, but the trend was then checked. Fixed investment rose from 9 per cent in 1950 to an average of 18 per cent in the period 1959–65, but then declined to only 9 per cent in 1977 and perhaps less in later years. While rising investment had been supported in the 1950s by growth in export earnings, in the early 1960s it depended on external commercial borrowing which could not be indefinitely sustained. Private consumption was about 72 per cent of the GDP estimate of 1950 and has been higher in most subsequent years; its relative size could be maintained even in 1959–65, the period of heavy investment, since there was then a succession of large import surpluses.

Table 6 shows estimates published by the World Bank of the uses of GDP in West African countries in 1960 and 1979. In this table, gross domestic investment comprises both fixed capital formation and changes in stocks. Gross domestic investment plus the resource balance equals gross domestic savings. Where the resource balance is negative, domestic investment is partly financed (in some cases, wholly financed) by an import surplus.

As might perhaps be expected, there are major inconsistencies (affecting Nigeria, Sierra Leone and Togo) between figures for 1979 in Table 6 and those for the same or an overlapping year derived from another source and given in Table 5. In particular, it will be observed that the ratio of government consumption in Sierra Leone is abnormally small in Table 5, but not in Table 6. Again, therefore, it is necessary to emphasize the very approximate character of these statistics, and their tendency to vary within the discretion of the responsible statisticians.

Table 6 Uses of GDP in West African countries, 1960 and 1979 (percentages of GDP estimates)

	Benin		*The Gambia*		*Ghana*		*Guinea*		*Guinea-Bissau*		*Ivory Coast*		*Liberia*	
	1960	*1979*	*1960*	*1979*	*1960*	*1979*	*1960*	*1979*	*1960*	*1979*	*1960*	*1979*	*1960*	*1979*
Private consumption	75	87	72	83	73	86	80	70	n/a	102	73	56	58	62
Government consumption	16	12	20	26	10	9	14	16	n/a		10	17	7	15
Gross domestic investment	15	21	13	22	24	5	5	15	n/a	32	15	31	28	27
Resource balance	−6	−20	−5	−31	−7	–	1	−1	n/a	−34	2	−4	7	−4
Gross domestic savings	9	1	8	−9	17	5	6	14	n/a	−2	17	27	35	23

The comparisons made in Table 6 are mostly consistent with the generalization that investment and government consumption have grown in relative importance. The ratio of government consumption is shown greater in 1979 than in 1960 for nearly all countries, and has fallen significantly only in Benin. In most countries, the investment rate is much higher in 1979; the most important exception is Ghana. Conversely, the ratio of private consumption has fallen in several countries, including the Ivory Coast and Nigeria; and elsewhere (Benin, The Gambia, Mali, Upper Volta) it appears to have been kept up only through growth in the negative resource balance.

Changes in the dependence of domestic expenditures on import surpluses have apparently been various. Thus, in Mali increases in all three ratios evidently required an increase in the import surplus sufficient to cover domestic dis-saving as well as investment. In Ghana, the negative resource balance of 1960 had disappeared by 1979 but domestic savings and investment had fallen to very low levels. In Guinea, domestic savings had roughly kept up with, and in Nigeria they had more than kept up with, considerable rises in the investment rate. In other countries, increased investment rates were accompanied by import surpluses, or relatively larger import

Mali		*Mauritania*		*Niger*		*Nigeria*		*Senegal*		*Sierra Leone*		*Togo*		*Upper Volta*	
1960	*1979*	*1960*	*1979*	*1960*	*1979*	*1960*	*1979*	*1960*	*1979*	*1960*	*1979*	*1960*	*1979*	*1960*	*1979*
79	82	79	47	79	72	87	58	68	} 98	n/a	78	88	74	94	89
12	23	24	39	9	9	6	10	17		n/a	18	8	15	10	14
14	15	37	51	13	28	13	31	16	21	n/a	15	11	39	10	24
−5	−20	−40	−37	−1	−9	−6	1	−1	−19	n/a	−11	−7	−28	−14	−27
9	−5	−3	14	12	19	7	32	15	2	n/a	4	4	11	−4	−3

Note: n/a indicates not available

Source: World Bank, *Accelerated Development in Sub-Saharan Africa* (1981), Annex Table 5.

surpluses, even where (as in the Ivory Coast, Niger and Togo) the rate of domestic saving had also risen.

Using the World Bank's estimates of the national GDPs in dollar terms as weights again, the average ratio of private consumption to GDP in West Africa in 1960 was 80 per cent, the ratio of government consumption 9 per cent and the investment rate 16 per cent. The negative resource balance averaged 5 per cent and gross domestic savings 11 per cent. On weighted average, the pattern in 1979 was very different. Private consumption now stood at 62 per cent of GDP, government consumption at 11 per cent, and investment at 28 per cent. The negative resource balance had fallen to 1 per cent and savings had risen to 27 per cent. This 1979 pattern corresponds with the weighted averages computed by the World Bank for all sixty 'middle-income' countries, save that in West Africa the percentages for investment and savings are a little higher and that for government consumption is a little lower.[16]

Like the changes in the overall sectoral origin of West African GDP in the period 1960 to 1979, this changing pattern of uses of GDP largely reflects structural changes in Nigeria and the increased economic importance of Nigeria in the West African region.

It should not be inferred from the estimates given in Tables 5 and 6 that an import surplus has been the invariable condition of West African countries, or of the region as a whole. The Gold Coast, Nigeria and Sierra Leone maintained large export surpluses in the decade following the Second World War, and accumulated substantial overseas assets as a result. The Ivory Coast appears to have earned export surpluses in every year but one between 1960 and 1977. Ghana, Sierra Leone and Togo had such surpluses in several years in the late 1960s and early 1970s. Large positive resource balances have been required in Liberia since the development of iron mining in the country, in order to finance recovery of foreign-owned capital and repayment of official debt, and were obtained until 1977. Finally, and most importantly, there were export surpluses in Nigeria throughout the period 1970 to 1976. There are other countries (Benin, Mali, Senegal, Upper Volta) for which import surpluses have been a regular condition, but the estimates for 1978 and 1979 cited above give a somewhat misleading impression of the extent to which West Africa as a whole has depended historically on foreign capital and inter-governmental largesse.

In conclusion of this section, some observations may be made concerning the composition of each of the three main categories of domestic expenditure. Within the aggregate of private consumption expenditure, the outstanding feature in West Africa is the high proportion attributed to food, beverages and tobacco. In the earliest estimates, these goods accounted for 76 per cent of private consumption in Nigeria in 1950 and 64 per cent in Ghana.[17] Later estimates – for Togo in 1972, Ghana in 1976 and Sierra Leone in 1977/78 – give them about 60 per cent.[18] In contrast, food, beverages and tobacco accounted for 24 per cent of private consumption in the UK in 1978 and 16 per cent in the USA. It must be added that, since in West Africa subsistence output is more important in food than in most other products, the proportion of *money-expenditure* devoted to food is likely to be appreciably smaller than the foregoing estimates suggest.

Among the purposes of government consumption, general administration, defence and education are of greatest importance, as indeed they are in economically advanced countries; together they usually account for about two-thirds of government consumption in West Africa.

Fixed investment can be broadly divided into building, other construction and land improvement on the one hand and machinery, vehicles and equipment on the other. There are considerable year-to-year fluctuations in the relative importance of these categories in West Africa, and it is difficult to discern trends or to establish contrasts among countries.

Factor shares in the GDPs

Estimates of GDP in West Africa have not usually been reached by computing factor incomes, and information is therefore more scarce on factor shares in the GDP than on its sectoral origin and uses. But in two countries, the earliest estimations did rely on the income method. In Ghana in the 1950s, recorded wages and salaries (later called the income from employment) accounted for about 15 to 25 per cent of the GDP at factor cost. Another 50 to 60 per cent was contributed by miscellaneous incomes and the incomes of cocoa farmers and brokers (later called the income from self-employment). The remainder, ranging between 15 and 35 per cent, consisted in the gross profits of companies, the operating surpluses of public boards and corporations, government income from property, and rent.[19] The greater part of this remainder was the operating surpluses of the official export monopoly, the Cocoa Marketing Board. Although the income from self-employment included elements of profits and rent, it would appear that the distribution of factor incomes was heavily weighted in favour of labour as opposed to property, and that this feature would have been even more pronounced but for the export monopoly, which usually held the producer price of cocoa well below its free-market value. Probably the relative shares of labour and property have changed little in Ghana since the 1950s.

Liberia in the 1960s is a contrasting case. Of the original GDP estimate for 1960, wages and salaries made up about 30 per cent, but other money-incomes earned by 'tribal households' only 4.5 per cent. Subsistence output contributed 11.5 per cent. The share of labour was therefore held not to exceed 46 per cent on even the most generous interpretation, and this was thought extraordinarily low; it was believed doubtful that there was any other country in which this share was less than one-half. (Further, over one-quarter of the wages and salaries, or nearly one-fifth of labour's share, was credited to non-African employees.) Of the remaining 54 per cent, nearly all was attributed to the gross profits of businesses and to rents, including especially the profits of the concessions in iron mining and rubber planting.[20]

Estimates of factor shares in the Liberian GDP continued to be published until 1969, and in that year the sum of wages and salaries, other money-incomes of tribal households and the imputed value of subsistence output was still under 50 per cent. This feature was less revealing of the character of Liberian politics, as was suggested at the time, than of the effects on the macroeconomic estimates for a small country of developing a large mining industry employing capital-intensive techniques and accordingly maintaining a high ratio of returns on capital (including depreciation) to value added. With the further development of West African mining in the 1970s, factor shares in other countries of the region, including especially Nigeria, must have shifted significantly toward the Liberian pattern.[21]

Domestic and national income

Computation of the GDP by the income method introduces the distinction between domestic and national income. The GDP is the earnings from producing goods and services within a country; domestic income is that quantity less the depreciation of capital. The Gross National Product (GNP) is the earnings from production *accruing to the residents* of a country; national income is that quantity less depreciation. The difference between the domestic and the national values is net factor payments abroad – i.e. payments to non-resident owners of factors of production used in the country, less receipts from the use of factors of production owned by residents but used outside the country.

The national income might alternatively be defined as the income accruing to the *nationals* of a country. The share of foreigners in the domestic income, whether or not they were resident and whether or not it was paid abroad, would be excluded from the national income. Though this conception of the national income does not appear out of keeping with the political ideology underlying national accounting, or with popular sentiment in some countries, it has not been favoured by national accountants.[22]

A third concept is national disposable income. This measurement is of national income (on the usual definition) less net current transfers abroad. The transfers may be between resident and non-resident households, including the remittances home of workers resident outside their own countries. They may also be inter-governmental, including grants to finance military outlays and such other current purposes as education, health care and feeding. Payments of contributions to international agencies, and grants from those agencies, would also appear in the current transfers between a country and the rest of the world. While factor payments abroad are made in return for productive services, current transfers are apparently unrequited.

Outflows of factor payments and current transfers from West Africa are partly attributable to foreign ownership, and the employment of non-African labour, in mining (as notably in Liberia, Mauritania and Nigeria) and, to a lesser extent, in large-scale manufacturing, trade and professional services (the Ivory Coast, Nigeria, Senegal) and plantation agriculture (Liberia). They also result from employment given to foreign Africans, as chiefly in the Ivory Coast, while on the other hand some countries (Upper Volta, Mali, Togo) receive considerable inflows of migrant workers' remittances.[23] Debt service also figures in factor payments abroad, while on the other hand some West African governments have at times received substantial earnings on investments held abroad. Excepting Nigeria and the Ivory Coast in recent years, West African countries are substantial net beneficiaries of inter-governmental transfers.

It appears possible to divide the West African countries into three groups with respect to the differences among domestic, national, and national disposable income. The first group is of countries with disposable incomes considerably larger than their domestic incomes, partly because, while having little use themselves for foreign factors of production, they have many nationals working abroad and remitting earnings home, and partly because their recurrent budgets are externally supported. In Upper Volta in 1974, for example, disposable income was about 125 per cent of domestic income at factor cost. Benin, Mali and Togo would belong with this group. So did Niger in 1969 when the last estimates were made; disposable income was then 110 per cent of domestic income.[24]

The second group is of countries where national income falls below the domestic by a relatively small margin and the difference is partly offset by a net inflow of current transfers. Thus, in Ghana in 1976, national was 1 per cent less than domestic income, but the difference was largely closed by current transfers. The pattern may be similar in Senegal. In Sierra Leone, the divergences are larger;

the estimates for 1978/79 show national as 95 per cent of domestic income, and disposable income as 97.5 per cent.

In the third group, national incomes fall, or have fallen, considerably short of the domestic. In Mauritania in 1973, the gap was 6 per cent, though it was more than offset by a net transfer inflow. In Nigeria, while net factor payments in 1958/59 had been trifling in relation to the GDP estimate, by 1970/71 they had risen to 7 per cent;[25] but this margin had diminished to less than 2 per cent by 1977/78. In the Ivory Coast, on the other hand, the divergences appear to have been growing steadily since the mid-1960s. The estimates for 1978 show national income as 94 per cent of the domestic, and disposable income as 89 per cent. Finally, the outstanding country in this group is Liberia. For 1960 the GNP was put at only 76 per cent of the GDP, and for 1964 at only 72 per cent. In 1968 it appeared as only 68 per cent if subsistence output were excluded. But in these estimates GNP was defined unusually to exclude all wage and salary earnings by non-African employees.[26] In an estimate for 1976, where possibly more orthodox definitions are being used, national income appears as 89 per cent of the domestic, and disposable income as 85 per cent.

Net outflows of factor payments and current transfers have the apparent effect of diminishing the disposable income of a country. Seemingly, that quantity would be greater if less use were made of foreign-owned factors of production, including migrant labour. In some West African countries, the apparent subtraction from the income available to residents (or to nationals) is undoubtedly large. And the outflow becomes even more important if it is expressed in relation, not to domestic income, but to savings and foreign exchange receipts, quantities usually considered to be crucial in determining economic growth. If, for example, the savings ratio is 20 per cent and foreign exchange receipts are 30 per cent of GDP, factor payments and transfers abroad equivalent to 10 per cent of GDP would absorb one-half of domestic savings and one-third of foreign exchange receipts. These large losses in key resources, attributable to the use of foreign factors of production, would appear to result in retarding progress in the countries experiencing them.

These considerations provide some support for measures, implemented to varying degrees in West African countries, such as controls on the employment of non-African labour, expulsion of foreign African workers, public participation in the ownership of foreign enterprises, reservation of some economic activities to nationals and the requirement that in others minimum levels must be observed of indigenous participation in the ownership of enterprises, and sundry other restraints on the activities of foreigners and the earnings they are allowed to remit abroad. No doubt governments also have other, and more pressing, reasons for discriminating in favour of their nationals and against foreigners, but estimation of losses of income, savings and foreign exchange plays a part. Further, the measures taken may actually check these losses to some extent. For, conceivably, the substitutability of national for foreign factors has been artificially obstructed or imperfectly recognized in the past. Perhaps, with more care, the services of some foreign factors can be obtained less expensively. Possibly to buy shares in foreign-owned enterprises is a better use of national savings than to create new capital assets.

Even so, it cannot be taken for granted that national disposable income will be absolutely greater if less use is made of foreign factors. On the contrary, it might reasonably be hypothesized that the foreign enterprise, technology, finance and skilled and unskilled labour are used in order to make good deficiencies in national resources and to overcome impediments to the exploitation of economic opportunities, and that these factors therefore add to national income at the same time as they secure their own reward. Similarly, the absolute amounts of savings and foreign exchange receipts are unlikely to be independent of the employment of foreign factors. Hence national savings, and foreign exchange receipts net of factor payments and transfers abroad, could be less in the absence of foreign factors. Further, the value of savings and foreign exchange for economic growth depends not only on their quantities but on the demand for them, or on how effectively they are used. The effective use may depend on the employment of foreign factors. Thus, economic growth (so far as it results from the supply of and the demand for savings and foreign exchange) may depend on this foreign participation no less than does the current level of national disposable income.

One might hazard the generalization that, while colonial prejudices led to excessive dependence on foreign factors of production in West Africa before independence, expectations of how readily those factors could be displaced without loss in national income or its rate of growth were also excessive in some countries.

Income distribution

The division of total income among factors of production has already been touched on. Other conceptions of income distribution currently excite more concern – the distribution among individuals, households or families, between urban and rural areas, among economic sectors and the administrative divisions or constituent parts of a country. The reason is an apprehension that economic growth in developing countries may be benefiting only better-off groups in the population (including, in West Africa, non-Africans), or benefiting those groups more than poorer groups. Growth in the GNP may therefore be accompanied by increasing inequality in its distribution. Such an outcome is variously represented as morally reprehensible, socially deleterious, politically hazardous, and inimical to a continuation of economic growth.

Where GNP and population estimates are available for the administrative or political divisions of a country, it is a simple matter to discover intranational disparities in income per head, with some regions or departments evidently more productive, or richer, than others. Sectoral disparities per employed person can be as readily calculated, where there are estimates of the sectoral distribution of the labour force. Average incomes in rural and urban areas can be contrasted by attributing agricultural output to rural residents and non-agricultural output to urban residents.[27]

These divergences are, of course, compatible with the observations made in Chapter 1, that the productiveness of labour, and therefore material levels of living, vary spatially in West Africa because of differences in climatic and other natural conditions, availability of commercial opportunities, and the location of mining and other productive investments. That the divergences may be accented by public policy (providing fiscal subsidies for some sectors, or affecting the location of investments) should not be allowed to obscure their more fundamental causation.

Inequalities in personal incomes (individual, household or family) are less simply documented. If all such incomes were known, their distribution could be presented pictorially as the graph of a frequency function or as a Lorenz curve, and the degree of inequality measured by a summary statistic such as the Gini coefficient.[28] The significance of this information would be in some doubt. Thus, the disparities revealed at any particular time are no measure of inequality over time, or in lifetime incomes. And it is not clear that relative inequality in personal incomes can be sensibly separated from the absolute levels of those incomes.

In any event, the information available on personal incomes in West African countries is fragmentary, of uncertain quality, and sometimes badly out of date. Reliance is placed on infrequent household budget surveys, and on wage and salary structures or income tax statistics which concern only small sections of the population. These shortcomings have not prevented calculation of Gini coefficients for West Africa, and of other measures of inequality such as the share of total income going to 'the lowest 40 per cent' or 'the top 20 per cent', nor the use of these measures for international comparison,[29] but they do strengthen doubts about the meaning of the results.

Apart from the practical difficulty of collecting information on personal incomes, there are conceptual problems in defining both the total available for distribution and the parties among which it is distributed. The relevant total might appear to be the GDP of a country less both its net factor payments abroad and its net current transfers abroad. But not all this total accrues to individuals, households or families. Some of it is received by corporate bodies as depreciation allowances and undistributed profits. Ought we therefore to be thinking of distribution among 'income-receiving units' rather than among more personal entities? Bizarre measurements of inequality could then result in small countries containing relatively large industries that were using capital-intensive techniques.[30] A better procedure would be to impute the retained earnings of corporate bodies to their owners in proportion to shareholdings, so far as these owners were both locally resident and private (the proviso is important, since governments and foreign interests are principal shareholders in many corporate enterprises in West Africa). A short-cut would be to assume that these locally and privately owned corporate earnings belong entirely to 'the rich' – to the top quintile or decile in the size-categories of income.

Government introduces further complications into the definition of income for purposes of measuring income distribution. The government also receives (or may be credited with) property income – through complete or partial ownership of corporate enterprises and through rights in natural resources. It is also in many West African countries the largest recipient of transfers from abroad. Potentially, therefore, it is another 'unit' receiving income. On what basis can this income be imputed to individuals, households or families? There are no shareholders in government.

Further, the government secures transfers to itself internally by its taxation of other recipients of income. Since these taxes are not paid in proportion to incomes, it would seem necessary to define incomes as after-tax magnitudes in measuring their interpersonal distribution. Questions of tax incidence then arise; it may not always be clear which incomes are being reduced by a particular tax.

Are these governmental receipts – from property, from international transfers, and from taxation – to be regarded as a subtraction from the total available for distribution among persons? The answer must obviously be not entirely, if the government in turn transfers part of its revenue to individuals as pensions and allowances. Personal cash-incomes are reduced by taxation, but some of them are increased by such transfer payments.

What of the government revenue that is spent on goods and services? Some part of it goes on purposes such as defence and justice that are arguably, if not evidently, of equal benefit to all members of the population. Other expenditures are on services or amenities creating benefits that are clearly not equally distributed. The benefits take the form of subsidies in consumption or production, with services, facilities or inputs being provided free or at less than cost. In some instances (feeder roads, water supply, communications), delivery is to communities, and the government does not control household participation in what has been provided; perhaps it can be assumed that participation is in proportion to income as otherwise constituted. In other instances (schooling, electricity, housing, farm inputs), benefits are traceable to persons, at least in principle.

Can such unequally distributed government services and amenities be disregarded on the ground that they provide income or inputs in kind and not in cash? That would seem hardly acceptable, since the inequality in distribution of services is often intentional and supposedly designed to 'correct' or 'improve' the distribution of income. Further, income is received in kind not only through government services but also by the subsistence activities of individuals, households and families. If the latter are to be included in measurements of income distribution (as they often are), why should not the former? In both cases, of course, intractable problems of money-valuation arise because the recipient is obtaining benefits he has not chosen to buy.

It may be added that even the distribution of cash-income is to some extent ambiguous since, as was observed in Chapter 1, the purchasing power of money varies in a country both socially and spatially.

To sum up, the income available for distribution can be defined as the GDP less net factor payments and current transfers abroad and less such government expenditures as are unallocable among persons. In principle, a share in this total would be the sum of cash-earnings less taxes incurred; any imputed share in retained corporate income; net transfers received from abroad; government transfers received; the imputed value of subsistence output produced; and the imputed value of government services received. In practice, few if any estimates of income distribution observe this principle.

Consider next the question whether the distribution should be measured among individuals, households or families. It is not difficult to imagine a situation in which a substantial proportion of the individuals in a population receive no income directly – especially if government services are left out of account. These individuals are dependants. Suppose they made up one-half of the population. Then the lower 50 per cent of individuals in the income distribution scale would receive no income, and the upper 50 per cent would receive the entire income.

It is unlikely that anyone would recognize that distribution as unequal. When distribution is said to be unequal, it is not on the ground that many individuals receive no income, but because incomes vary among those who do receive them. It would therefore appear more appropriate to measure distribution among individual income-recipients than among all individuals. Suppose, then, that the upper half of income-recipients receives two-thirds of the total income, and the lower half one-third. Apparently income is here unequally distributed; individuals in the upper half have twice as much on average as those in the lower half. But suppose also that the individuals in the upper half have twice as many dependants on average as those in the lower half. Then incomes averaged over the people they support would be equal between the groups of income-recipients. Inequality in the distribution among income-recipients could be a condition of equality in the distribution among all beneficiaries.

The same conclusion could be reached if the distribution were measured among households rather than among individual income-recipients. Households might be thought a more appropriate unit, since two or more individual incomes may be pooled in support of a group of people, and since some incomes (for example, those derived from subsistence agriculture) may be deemed to accrue to households rather than to individuals. Because households are not of uniform size, a household distribution averaged per head will be different from the household distribution itself. The measured inequality could, again, be a condition of equality per head.

These possibilities are more than merely theoretical. It is to be expected for most individuals that income and the number of persons supported by it will move together. Thus, as a man matures and becomes more productive, he is likely to acquire a wife (or more wives) and children. Income and the number of dependants move together. Similarly, household income and household size are correlated. Better-off households tend to be bigger, partly because they contain more income-recipients, partly through procreation, and partly because they attract new members having claims on the household head. Hence both individual and household incomes tend to be positively related with the number of persons they support. It follows that distributions averaged among beneficiaries not only may be, but almost certainly are, less unequal than the distributions among individual income-recipients or among households.

It is another question, of course, whether income is distributed equitably within a household or among a circle of dependants. An averaging per head does not imply that everyone concerned receives this average, or as much of the total as he or she may be deemed to deserve. Studies of income distribution usually stop short of intra-household distribution – the transfers from those who earn or control household income to those who are dependent. At least in part, the reason is ideological: intra-household distribution is generally believed to be an area unsuitable for public intervention and hence for statistical investigation.

Whether the family corresponds with the household depends on how the family is defined. Defined as parents and their children, it could often be less than the household. Defined extensively, as it is in West Africa, it is so much larger and less readily identifiable than the household as not to be a feasible unit for measuring income distribution. Consequently, in addition to the intra-household transfers lying outside conventional measurements of distribution, there are also intra-familial transfers. A step beyond the latter lie intra-communal transfers, as when a migrant community remits funds to finance improvements at its place of origin. That such informal transfers, unlike government transfers, are excluded from measurement does not, of course, negate their practical effect, whether this be to increase or (as might appear more likely) to reduce inequality.

The Gini coefficients of income inequality which have been calculated for Dahomey (1959), the Ivory Coast (1959 and 1970), Nigeria (1960 and 1967), Senegal (1960) and Sierra Leone (1968) are in a relatively high range of about 0.45 to 0.60; only for Niger (1959) has a fairly low degree of inequality been shown, with a coefficient of 0.34. For Nigeria in the mid-1970s, a coefficient as high as 0.7 or 0.8 has been suggested.[31] The share in total income of 'the lowest 40 per cent' appears as one-tenth or less in the Ivory Coast,[32] Nigeria, Senegal and Sierra Leone. Uncertainties about how income and the units sharing in it have been defined make it difficult to know how seriously to take these figures. The factual basis of the estimates is also sometimes unclear; some figures appear to have been produced (and they are then frequently cited) on a very slender basis indeed.[33]

These estimates notwithstanding, there are some forces telling against high interpersonal inequality in West Africa. Agriculture is the source of livelihoods for most people. The ratio of population to cultivable land is favourable in many parts; access to the land is therefore free or cheap for many people; there are very large numbers of agricultural enterprises and the typical scale of operation is small. Variations in agricultural earnings depend more on differences in natural conditions, household sizes and access to markets than on property rights. Where agricultural labour is hired, much of it is labour that has migrated from less favoured areas, and its use restrains differences in earnings. The freedom of labour to migrate also helps limit differences between rural and urban earnings, and the labour-intensive character of most urban activities (public administration, building, small-scale manufacturing, distribution and other services) produces a concentration of employment, both formalized and informal, at the lower end of the range of remuneration. Finally, the extensive nature of family obligations in West Africa leads to much transference of income from better-off to worse-off individuals.

Yet relatively small differences in incomes among the bulk of the labour force or population are compatible with high measures of inequality overall. It appears that, in many developing countries, differences in the earnings of labour are the principal factor making for inequality in incomes.[34] In West Africa, scarcity of technically trained and highly educated labour has led to payment of high differentials in the labour market, a matter discussed further in Chapter 3. The coexistence of a large number of workers, whose remuneration was fairly uniform, and a small number whose pay was relatively very high would produce high measurements of inequality.

The dispersion of earnings from labour has possibly increased in West Africa in recent times with the development of new and highly capitalized mining, manufacturing and service industries. Other forces may have been making for greater inequality. The returns from ownership of urban property have grown rapidly as urbanization has accelerated. Inflation has redistributed income from wage and salary earners to recipients of profits and rents. Direct taxation, as it has been extended, has probably not been progressive in its incidence. Growing government regulation of economic life has provided increasing opportunities of enrichment for persons or groups with political power or influence. Kinship obligations are perhaps taken less seriously than they used to be.

What greater inequality signifies to those who experience it is debatable. Certainly, the view that there is widespread antipathy to economic inequality in developing areas like West Africa should be questioned. Inequality could deepen because of redistribution of a given income in favour of richer people. In this case, some groups would be impoverished both relatively and absolutely. But inequality could also increase because some incomes rose while others did not, or because incomes rose at unequal rates. In this case, some groups would be relatively poorer, but absolutely no worse off, or even better off. It is likely that people in a poor region like West Africa attach more importance to absolute than to relative changes in their material condition, and that therefore gains in the incomes of some groups would not be regarded as seriously depreciated by an accompanying increase in the degree of inequality, so long as they did not imply absolute reductions in other incomes. For a time, at least, such unequally distributed gains might even be welcomed by those who were relatively disadvantaged by them, as holding out the promise that they too could benefit in the future.[35]

Income distribution and the composition of demand

However imperfectly we can measure it, and however uncertain its political significance, the distribution of income does help to determine the composition of demand and the allocation of resources in response to that demand. A less unequal distribution would mean less demand for the goods and services favoured by high income-earners and well-off households, and more demand for the products desired by the poor.

Several advantages have been claimed for such a shift in the composition of demand. The low-income goods are postulated to be products (like foodstuffs, traditional housing and the cheaper varieties of simple manufactures) in which there is a local comparative advantage. Hence there would be more demand for local production and less for imports – whether imports of finished high-income goods or of the inputs required for local production of such goods. Additions to aggregate demand would consequently have a stronger local multiplier effect. More employment would be given locally, not only because of the shift in demand toward local products but also because the low-income goods are postulated to be those produced by relatively simple technologies with high labour coefficients. The lesser use of advanced and complicated technologies would imply less need of foreign factors of production, and hence reduction in factor payments and transfers abroad. Reductions in those factor payments and transfers and in import demand would lessen the dependence of local economic activities on receipts of foreign exchange. Finally, there would be produced a pattern of consumers' tastes, and a technological capacity for satisfying them, more appropriate to modal living conditions and local resources.

Quite apart, therefore, from ethical and political arguments in favour of reduction in income inequality, that process has been justified on the grounds that it would create more employment locally, reduce dependence on external markets and finance, and produce greater harmony between economic aspirations and local resources.

A rival view that high inequality is economically desirable as a basis for high rates of saving and capital formation has lost favour.

Raising the investment rate has been a less formidable difficulty in many developing countries than was expected (see Table 6), and it is argued that governments can attract or enforce savings more reliably than inequality can engender them.

Perhaps so, but the economic arguments in favour of lessening inequality are also open to criticism with reference to West Africa. In the first place, it is not certain that the tastes of consumers vary so much among income-classes that redistribution of income would make a pronounced change in the composition of demand. In particular, the marginal propensity to import may be high for all income-groups, if local productive capacity is unable efficiently to satisfy more than elemental wants. For example, the taste for imported foodstuffs such as rice, wheat and canned fish is not confined to the rich.

Second, government measures to reduce inequality are likely in practice to consist largely in spending on government services and amenities. It is not obvious that such spending is concentrated on goods and services in the production of which there is local comparative advantage and labour-intensive technologies are used. For example, those features would not be characteristic of much spending on construction, telecommunications, health care and fuel subsidies. The government too may have a high propensity to import.

Third, supposing that demand were diverted from imports to home production, the effects on employment would depend on the elasticities of factor supply. Even for labour, a high elasticity of supply cannot be taken for granted, since much of the unemployment observable in West African towns expresses a preference for employments other than those that are readily available; it is not apparent, for instance, that this unemployment would be reduced by a rise in market demand for local foodstuffs. The combining of labour with other factors to create the capacity to respond to additional demand is likely to be still more constrained; for example, new industrial ventures in West Africa have frequently been impeded by insufficient, or insufficiently reliable, supplies of local raw materials. It is, of course, precisely because of the limited adaptability of local productive structures that the increase of income (and of employment) comes to depend heavily on exports and perhaps also on the use of foreign factors, and that the import ratio becomes large.

Fourth, a strong dependence of local economic activities on

receipts of foreign exchange is not self-evidently an economic disadvantage, if it means that the value of those activities is greater than would otherwise be attained.

Finally, the harmonizing of consumers' tastes and producers' technologies with local circumstances is a prescription of quietude. It would remove the pressures for economic improvement that arise from inappropriate aspirations, including the aspiration, powerful in West Africa, to make oneself and one's kin better off than the majority.

These criticisms suggest that redistribution of income would do little to reduce the economic 'openness' (or high dependence on external trade) of West African countries, except by diminishing the productiveness of their resources. If local productive capacity can be made more various, versatile and efficient, demand can be expected to adjust, since the choice between national and foreign goods and factors is primarily economic. But expectations that improvements in productiveness will follow simply through changing the composition of demand, either by redistribution or in any other way, are likely to be disappointed; such, for example, has been the experience of import controls in several West African countries.

Instability

The economic openness of West African countries makes their economic life vulnerable to changes external to themselves. For example, the proceeds may fall from Ghanaian or Ivorian cocoa because world supply has risen relatively to the demand for this commodity; or from Liberian iron ore because of industrial recession in the countries importing the mineral; or from Nigerian oil because the Nigerian government has overestimated the price at which the world market would absorb the quantity it wishes to be sold. There may also, of course, be upward fluctuations in export proceeds attributable to changes in external market conditions. Again, the costs of a major import, such as rice or fuel oil, may be volatile because of shifts in the balance of world supply and demand. Local demand may also be strongly affected by a large inflow of direct foreign investment, such as Liberia experienced in the early 1960s – and again by the later dwindling of the inflow. The instability arising from such causes tends to be more pronounced in West Africa than in some other countries with no less considerable ratios of foreign trade because of the less diversified nature of the West African trade and the size which foreign investment may assume relatively to the total economic activity of small countries.

Turning first to export instability, a feature in West Africa is that its impact is primarily on governments and foreign investors rather than local producers and employees. The exports are mostly of crops and minerals. The arrangements for exporting the major crops are such that the prices paid to producers and the margins obtained by local traders are fixed from season to season. It does not follow that the proceeds from growing and collecting a crop are pre-determined, for they will vary with the size of the crop. But it does follow that producers and local traders are insulated, at least for the duration of a season, from fluctuations in the export price, the impact of which will be entirely on the trading surpluses accruing to the official marketing board or stabilization fund which guarantees the local price, and on the proceeds from government export duties where they are related to the export price.

In the case of large-scale mining, the initial impact of export instability is on company profits, and hence partly on factor payments and transfers abroad (so far as a company remits profits to foreign owners) and partly on public receipts (so far as the government taxes profits or shares in them through public ownership). Other parties may be affected later, if a company is induced by the external change to contract or expand its production, but it is through the public share in profits that local vulnerability is immediately felt.

As to instability in import prices, this is felt less in West Africa than in countries depending heavily on imports of primary products. Large proportions of the import expenditures of West African countries are on manufactures and services, whose prices are less changeable. The more volatile sectors of West African imports are fuel and food. It might be thought that the effect of changes in the prices of these import commodities would be widely felt. But even here government finances may bear the brunt of upward price changes, if the government elects to protect consumers by subsidizing local prices. Thus, rice has been subsidized in countries (Liberia, Mali, Sierra Leone) importing large quantities of that foodstuff, and in other countries (Nigeria, Ghana) the local prices of petroleum products were restrained by reductions in tax rates during the oil price rises of the 1970s.

The instability of foreign direct investment may make itself felt in fluctuations in demand for labour, especially in mining where investment is substantially in construction. This source of instability stems from irregularity in the occurrence of new profit-making opportunities, and it is not evident that the instability would be less if the investments were being made by nationals instead of foreigners.

In principle, at least, there are several ways of controlling the effects of these external changes. Thus, the statutory monopolies for crop exporting and the stabilization funds were at one time represented to be the means of maintaining local incomes in face of export instability; the surpluses they obtained in times of high export prices were to be reserves that could be drawn on when export prices were less favourable. In the event, these institutions became absorbed in the financing of government (see Chapter 5), and the burden of instability (or, more accurately, of an unstable export price) was merely transferred from producers to the governments. West African governments consequently became supporters of proposals for controlling by international agreement the range within which prices of their exports might fluctuate, and some of them have been parties to the International Coffee and Cocoa Agreements. The governments have also been able to draw, subject to certain conditions, on stabilization or compensatory loan funds maintained by the European Economic Community under the conventions of Lomé (see Chapter 4) and by the International Monetary Fund. In addition, they may be able to borrow abroad commercially to cover shortfalls in their export receipts. As to the fluctuations in foreign direct investment, they might be offset by variations in public investment, assuming that the government has funds available or is able, again, to borrow abroad, and that it has worthwhile schemes to execute.

That these solutions may not always be available in practice is perhaps less important than that the problem of instability has not been a matter of overriding concern to West African governments. Indeed, in some countries, government policies have probably been more serious causes of instability than have fluctuations in export or import prices or foreign direct investment. Thus, in Ghana the monopolizing of cocoa exports has prevented the local producer price from moving inversely with the size of the crop, as would have happened in a free market because of the large share of Ghanaian output in world supply; the year-to-year fluctuations in cocoa farmers' aggregate receipts have therefore probably been greater than they would have been in the absence of the Marketing Board. In several countries including Ghana, Sierra Leone, Mali and Mauritania, external borrowing to finance public investments, and the obligation to service these borrowings, would appear to have occasioned more instability than have changes in the inflow of foreign direct investment. Demand has also been destabilized by the rapid running down of official external reserves as a result of sudden expansions in government development programmes, as in Ghana in 1959–61, Sierra Leone in 1965–6 and Nigeria in 1976–8. These disturbances have been aggravated by deficit financing in countries possessing autonomous monetary systems; Ghana and Sierra Leone have been leading examples. Finally, a major source of instability in import expenditure in several countries has been stock accumulation by speculators anticipating changes in government policy toward importation.

One reason why stability in incomes, employment and prices has been neglected as a policy objective is that West African governments have attached much more importance to economic growth. Their concern has been for a larger rather than a more secure national product. Thus, even the interest in international commodity agreements is less a desire for more stable export prices than a hope that the agreements may be the means of raising export prices relatively to import prices. Policies of economic growth and development in West Africa and the means of executing them will be discussed in Chapters 5 and 6. Since much of the present chapter has been devoted to examination of aggregative measurements of economic life in the region, it is appropriate to point out now the shortcomings of measured rates of economic growth.

Economic growth rates

If the GDP (or GNP) is accepted as a valuation of economic activity during a year, the change in this quantity between one year and the next naturally suggests itself as the overall measurement of a country's economic progress. Since the quantities are expressed in money, they must be corrected to remove the effects of any general change in the value of money. Thus, the GDP of the second year may be arrived at by valuing the volumes of net output in that year at the prices obtaining in the first year. The resulting 'real' or constant-price change, usually calculated as a percentage to one decimal place, seems to be regarded by national policy-makers and their economic advisers as an indicator of profound significance. For longer periods, compound annual rates of real growth can be calculated. International comparisons of such rates are often made to show the progress of countries, or even of large groupings of countries, relatively to one another. Domestic or national products are also divided by annual estimates of population to obtain output per head, the real growth rate in which is often regarded as a truer measure of progress, on the ground that some growth in the aggregate product is achieved only by (or is in some sense offset by) increase in the number of people.

Table 7 shows estimates, drawn from several World Bank publications, of the compound annual rates of real growth in GNP per head in West African countries for several time periods beginning 1950. It would appear that the growth rate in this quantity was considerable, over the entire thirty-year period from 1950, in Nigeria and the Ivory Coast, which are economically the two largest

Table 7 Estimates of annual percentage rates of real growth in GNP per head

	(1) *1950–60*	*(2)* *1960–70*	*(3)* *1970–8*	*(4)* *1960–79*
Benin	n/a	0.8	1.4	0.6
The Gambia	−0.2	3.8	2.9	2.6
Ghana	1.9	−0.7	−3.0	−0.8
Guinea	n/a	0.0	0.8	0.3
Guinea-Bissau	n/a	n/a	2.9	n/a
Ivory Coast	0.0	4.3	0.9	2.4
Liberia	n/a	1.1	0.2	1.6
Mali	1.2	1.2	1.8	1.1
Mauritania	−2.6	5.6	−0.6	1.9
Niger	n/a	−0.5	−0.6	−1.3
Nigeria	2.1	0.4	4.4	3.7
Senegal	4.4	−1.6	−0.3	−0.2
Sierra Leone	n/a	0.0	−1.3	0.4
Togo	−1.4	5.7	1.7	3.6
Upper Volta	−0.7	0.6	−1.0	0.3

Note: n/a indicates not available.

Sources: Cols. (1) and (2) from David Morawetz, *Twenty-Five Years of Economic Development 1950 to 1975* (Washington DC: World Bank, 1977), Statistical Appx, Table A1, pp. 77–8; col. (3) from *1980 World Bank Atlas*, p. 5; col. (4) from *Accelerated Development in Sub-Saharan Africa* (Washington DC: World Bank, 1981), Statistical Annex, Table 1.

countries, and also in Togo and The Gambia. As to the shorter time periods, the growth rate appears to have been fastest in the 1950s in Senegal, Nigeria and Ghana; in the 1960s in Togo, Mauritania, the Ivory Coast and The Gambia; and in the 1970s in Nigeria, The Gambia and Guinea-Bissau. There are also many instances of retrogression, seemingly most pronounced in Mauritania in the 1950s, Senegal in the 1960s, and Ghana in the 1970s.

Since population growth has been rapid, possibly 2 to 2.5 per cent annually in the region as a whole, growth rates in GNP are much higher than those estimated for GNP per head. In Nigeria, in the twenty years from 1960, the GNP growth rate averages over 6 per cent, and in the Ivory Coast over 7 per cent per year; in West Africa as a whole (weighting the GNPs by their 1960 dollar values), it is about 4 per cent.

While the impressions of real changes given by these figures are probably not seriously misleading, some doubts must be expressed concerning both the meaning of economic growth rates and their accuracy as measurements of changes in whatever they do mean. These reservations concerning the validity of inter-temporal comparisons of GNP are analogous to those made about international comparisons in Chapter 1.

Economic growth rates are usually interpreted as measurements of changes in the productive power of a country, or the aggregate (or average) wellbeing of its population, or both. These interpretations depend on the definitions adopted of final economic goods and services, the weighting by prices of these products as means of adding them up, and the assumption that human wellbeing is positively related with (if not actually measured by) changes in the resulting totals. The distinctions between economic and non-economic activities, and between final and intermediate goods and services, are arbitrary: the former because some subsistence activities are counted as economic and included in the measurement while others are not, the latter because the determination of finality in chains of production can be only conventional. Two illustrations will show why this arbitrariness matters in the measurement of growth. First, if more cooked food is purchased and less prepared at home, output included in the economic measurement has risen only because output excluded has fallen. Second, the use of more public transport to move workers to and from their places of employment would be conventionally counted as an addition to final output, yet it might more reasonably be regarded as a cost incurred through increasing concentration of production. In short, the validity of the growth rate is put in doubt by changes over time in the division of resources between economic and non-economic uses. The necessity to assign prices to output increases this uncertainty. It cannot be assumed that these money-values measure the relative marginal satisfactions derived from using various goods and services, or the relative marginal costs of producing them, when some of the values are merely imputed to non-marketed production, others belong to government output which users or beneficiaries do not choose to buy, and others are affected by imperfections of competition among producers. Nor can it be assumed that the same money-value represents equal satisfaction to all consumers, when incomes and tastes vary. If the meaning of relative prices is in doubt, so must be the meaning of increase in a total of output valued at these prices and changing in composition over time. Finally, to connect growth in an aggregate of economic production with improvement in social wellbeing abstracts both from the interpersonal distribution of the production and from cultural changes which may be inseparable from the growth of production and are not less a part of human wellbeing.[36]

All that the growth in GNP measures, therefore, is growth in what the statisticians have chosen to count as GNP, valued at the money-costs of producing it rather than by the benefits derived from it. Even this measurement, for whatever it may be worth, is statistically unreliable. Some indication of this unreliability is present in Table 7, where all the figures come from the same authority but there are obvious inconsistencies between column (4), relating to 1960–79, and columns (2) and (3), relating to 1960–70 and 1970–8, so far as Benin, The Gambia, Liberia, Niger, Senegal and Sierra Leone are concerned.

Even more strikingly inconsistent are the estimates shown in Table 8, which have been made by two French agencies and four international organizations, of the annual real growth rate of GDP in the 1960s in five West African countries. The disagreement among these estimates is such that the highest exceeds the lowest by a factor of two in Senegal, five in Upper Volta and Niger, and no less than thirteen in Mali. Only the figures for Mauritania are fairly consistent, and even here the maximum exceeds the minimum by nearly one-quarter. It will be noticed, too, that none of these estimates agrees with those shown in column (2) of Table 7; for Niger and Senegal, not even the sign of the change (positive or negative) agrees.

Each of these conflicting estimates is afflicted by the margins of error in whatever GDP estimates have been used for the first and last years of the period. In Chapter 1, a suggestion was cited that the margin of error in African GDP estimates averaged some 20 per cent. Applying this figure to (for example) the United Nations estimate for Niger in Table 8, the compound growth rate reported as 4.7 per cent could apparently lie anywhere in a range extending from 0.5 to 9.0 per cent.

Table 8 Estimates of annual percentage rates of real growth of GDP in five Sahelian countries, 1960–70

	SOEC	*SIEC*	*UN*	*OECD*	*IBRD*	*UNCTAD*
Mali	3.0	2.5	0.5	2.8	6.6	5.2
Mauritania	7.4	8.0	7.7	7.3	6.5	6.9
Niger	2.4	2.0	4.7	2.0	0.9	2.4
Senegal	1.6	1.6	1.3	2.0	2.1	1.0
Upper Volta	3.9	3.3	3.0	2.0	1.5	0.7

Notes: SOEC is Secteur des Etudes Socio-Economiques de Synthèse and SIEC is Secteur Information Economique et Conjoncture, both Bureau des Programmes, Direction de l'Aide au Développement, Ministère de la Coopération, Paris; UN is United Nations; OECD the Development Assistance Committee of the Organization for Economic Cooperation and Development; IBRD the World Bank; and UNCTAD the United Nations Commission for Trade and Development.

Source: Economie, Emploi et Formation: Evolution et Perspectives pour 14 Etats Africains et Malgache, 1, *Evolution du P.I.B. 1950–1970, Perspectives 1970–1990* (Paris: Direction de l'Aide au Développement, Ministère de la Coopération, 1974), cited in Elliot Berg, *The Recent Economic Evolution of the Sahel* (Ann Arbor, Michigan: Center for Research on Economic Development, University of Michigan, 1975), p. 11.

The effect on a growth rate of any assumed error in the underlying aggregates must increase as the period under consideration is shortened. On the other hand, it can be argued that much the same errors may be made in computing the estimates for adjacent years, so that the difference between these estimates is much less unreliable than the estimates themselves. But if, on this ground, the assumed margin of error in the estimates of adjacent years were reduced to only 2 per cent, the one-year growth rate would remain just as uncertain as the ten-year growth rate in the previous illustration using a 20 per cent error. As Oskar Morgenstern observed of GDP figures, 'a reliable growth rate of two significant digits is impossible to establish', and 'even the first digit is in grave doubt', albeit the emphasis of public discussion is on the second digit if not the first decimal.[37]

Statistically as well as conceptually, therefore, measurements of the rate of increase in GDP or GNP, or in these quantities per head of population, are of dubious value in describing a changing reality – and not only in West Africa. This is not to deny the great importance of economic growth in practice, nor its significance as a subject of enquiry, nor that changes in many economic quantities are measurable. But it is chimerical to essay an overall measurement of growth in the economic life of a country, and facile to suppose that such a measurement adequately records improvement in the human condition.

Summary

This chapter has shown that the structures of West African economies are very imperfectly described by the available estimates of aggregate production, expenditure and income. Among productive sectors, estimates of output may contain margins of error of 10 per cent at best and 50 per cent at worst. Among expenditures, private consumption – often the greater part of the GDP – is a residual containing much roughly estimated and arbitrarily valued subsistence activity. Little information is available on factoral shares in income. GDP estimates obtained from different sources, or from the same source at different times, are characteristically inconsistent. The totals could well be in error, on average, by 20 per cent in either direction. It would not be surprising if there was a tendency to underestimate the scale of non-formalized activities about which little is known with certainty, and conversely to exaggerate the importance of formalized and enumerated activities.

As compared with industrially advanced countries, West African countries appear to derive large proportions (one-quarter to one-half) of their GDPs from primary activities, and small proportions from manufacturing. Among West African countries, some of the contrasts shown by the estimates are undoubtedly spurious, but there are real differences in the relative importance of mining and construction. Contraction in the relative importance of agriculture is shown for several countries in the twenty years beginning 1960, and is attributable mainly to the further development of mining and the continuing commercialization of economic life. For the region as a whole, agriculture appears to have fallen from over 50 per cent of the combined GDP estimate in 1960 to less than 30 per cent in 1979, while industry (mining, manufacturing, construction and public utilities) has expanded from 15 to nearly 40 per cent. These changes reflect the changing structure of the Nigerian GDP estimate, following the exploitation of oil deposits, and the growing share of Nigeria in the combined West African GDP (from about 45 per cent in 1960 to 70 per cent in 1979).

The undeveloped character of input–output relations in West African production reflects the current comparative advantage of the region in international trade.

The ratios to GDP of the main categories of final expenditure differ considerably among countries of the region, and they may fluctuate sharply with changes in the resource balance. Over time, there has been a tendency in most countries for the ratios of fixed investment and government consumption to rise, and for the ratio of private consumption to fall – though it has been sometimes held up by a negative resource balance. For West Africa as a whole, private consumption between 1960 and 1979 may have fallen from about 80 per cent to not much more than 60 per cent, while fixed investment appears to have risen from 16 to 28 per cent and domestic savings from 11 to 27 per cent. The pattern of use of GDP for the region as a whole in 1979 is similar to that estimated for all 'middle income' countries on average; the economic predominance of Nigeria in the region is, again, reflected.

Analysis of the GDP by factor shares is not usually available, but estimates for Ghana in the 1950s and Liberia in the 1960s show a marked contrast in the division of income between labour and property. With the further development of mining in the region, other countries, including Nigeria, must have shifted toward the Liberian pattern.

Some of the countries in the region, such as Upper Volta, dispose of more income nationally than they produce domestically. In others, like the Ivory Coast and Liberia, there are large net outflows of factor payments and current transfers, attributable to their employment of foreign-owned capital and immigrant labour. Displacement of foreign-owned by locally-owned factors of production might appear an obvious means of increasing national disposable income, and more particularly national savings and disposable foreign exchange earnings, but it cannot be taken for granted that this substitution is practicable without losing output or retarding its growth.

The distribution of income currently excites much concern on the ground that economic growth may be accompanied by increasing inequality. Differences in average output per head among economic sectors or administrative divisions of a country can be readily demonstrated; for the most part, they depend on natural conditions and sectoral differences in investment. Inequality in personal incomes is less simply documented. The available information is fragmentary, of uncertain quality, and sometimes badly out of date. Important questions arise in the definition both of income and of the units among which its distribution is to be measured, and it is not always clear how these questions have been answered. The high measurements of inequality (Gini coefficients ranging mostly from 0.45 to 0.6 or even more) which have been reported in West Africa must therefore be regarded with considerable reserve. So far as they are valid, the main reason is likely to be the coexistence of numerous workers with fairly uniform earnings and a small number of highly educated and skilled workers receiving very large differentials. The political significance of inequality is not as straightforward as is sometimes suggested; static measurements tell one nothing of lifetime opportunities or of absolute changes in income over time, and it may be misleading to believe that egalitarianism is a popular cause in West Africa.

It is questionable whether a less unequal distribution of income would do much to reduce the economic 'openness' (or high dependence on external trade) of West African countries, except by diminishing the productiveness of their resources.

The initial impact of instability in export prices in West Africa is on government revenues and possibly on factor payments and transfers abroad. Government finances may also be affected by instability in import prices where the prices of key import commodities, such as rice and mineral fuel, are subsidized. The effects of these external disturbances can, in principle, be controlled in various ways – through stabilization funds, international agreements or external borrowing. In practice, some West African governments have created more economic instability than they have offset. One explanation is that their objectives have lain in growth rather than in stability.

Finally, estimates of GDP at constant prices indicate a brisk rate of economic growth in the West African region in the thirty years beginning 1950, much of it attributable to the two economically largest countries, Nigeria and the Ivory Coast. But there is little agreement on what the growth rates are for particular countries over particular periods, and the rates reported must be deemed to contain margins of error of several percentage points. The validity of these growth rates as measurements of improvement in social welfare, or even in economic performance, is also open to question.

3

Population and Labour Force

Demographic data

Population censuses were taken during the 1970s in all the West African countries except Guinea and Benin (where demographic data rest on surveys conducted as long ago as 1954–5 and 1961 respectively). In the francophone countries, they were the first censuses, although they had been preceded by demographic surveys. Elsewhere, there had been censuses before the 1970s. Most notably, Ghana has had three since the Second World War – in 1948, 1960 and 1970. The Gambia had censuses in 1963 and 1973; Guinea-Bissau in 1950, 1960, 1970 and 1979; Liberia in 1962 and 1974; and Sierra Leone in 1963 and 1974. Nigeria had censuses in 1952–3, 1962, 1963 and 1973, but the second and the last of this series were annulled.

Census-taking in Africa is hampered by a need for personal contact between enumerators and those who are being enumerated, and by shortage of trained enumerators; hence counting has sometimes extended over weeks or even months. It is generally believed that administrative shortcomings, along with some popular aversion to enumeration, produced substantial under-counting and other misinformation, especially in the earlier censuses. Thus, the 1948 census total in Ghana was thought to be an under-count by about 7.5 per cent in the light of the 1960 results. The 1960 total has itself been adjusted upward by 2.5 per cent, and that for 1970 by 1.6 per cent.[1] It might be supposed, as this illustration suggests, that the accuracy of successive censuses would grow as experience was gained, administrative resources increased, and the public became more cooperative. But it is noteworthy that in Sierra Leone, while the 1963 census had been thought to under-enumerate the population by 5 per cent, an addition of 10 per cent was made to the enumerated total of 1974 to produce the official population estimate. Under-enumeration by 11 per cent was officially assumed in the Liberian census of 1974. In Nigeria, a presumably under-enumerated population in 1952–3 was replaced as a basis for demographic projections by an almost certainly over-enumerated population in 1963. Everyone is agreed that demographic data in West Africa are mostly of poor quality. Whether they are getting better is a moot point.

Since Nigerians are probably more than one-half of all West Africans, some explanation must be given of why their number is so uncertain. The census of 1952–3 produced a total of 30.42 million, of whom 16.84 million (55 per cent) were attributed to northern Nigeria. In the south, there was a popular belief that the colonial authorities had inflated the northern total in order to maintain a majority of northern representation in the Federal House of Assembly; there was an expectation too that this deception would be unmasked in the census taken in May 1962. Though the results of the 1962 census were never published, they are understood to have shown a national total of about 45 million, of whom 22 million, less than half, were enumerated in the Northern Region. The Federal Census Officer then reported that the figures recorded in the greater part of the Eastern Region were false and had been inflated, and that results were available for only a few of the census districts in the Western Region because of weaknesses in organization. Subsequently, the total for the North, which had been the first declared, was revised or 'verified' as 31 million. The census of 1962 had to be acknowledged a failure, and a new count was arranged in November 1963.

Extraordinary measures were taken in this recount in an effort to ensure accuracy, but it was perhaps to be expected that none of the three Regional governments, nor even any census district, would admit its earlier figure to have been inflated, as it would have done by returning now a lower figure. The effect of recounting was in fact further to increase the national total. It was eventually put at 55.67 million, within which the Northern total was 29.8 million (53.5 per cent). The East's figure of 12.4 million was little different from its total of 1962 that had been declared false and inflated. It would seem that the national total for 1963 was produced by the determination of the Northern and Western governments that they would match the East's capacity for falsification. The final results implied that between 1953 and 1963 the national population increased at an annual average rate of 6.3 per cent, and that of the Western Region by 8.3 per cent annually. Under-counting in 1952–

3 could explain away some of this apparent demographic absurdity, but hardly all of it.

The Nigerian censuses of 1962 and 1963 can be interpreted as an attempt by the Eastern Regional government to remove the northern majority in the Federal Assembly and a successful riposte by the North. But underlying the gerrymandering was an enthusiasm for enumeration which had not been present in 1952–3 and which depended not only on local support for regional chauvinism but also on widespread acceptance of an association between the sizes of communities and the geographical distribution of public expenditure on services and amenities. Indeed, the politicians had publicized the census of 1962 as the means both of determining political representation and of establishing claims to public services, overhead capital and the siting of industrial plants. The response to this salesmanship perhaps exceeded expectations.[2]

The civilian governments of the federation were overthrown in 1966 and there began a period of military rule which included the civil war of 1967–70 and continued until 1979. An item in the federal military government's programme for returning to civilian rule, announced soon after the end of the civil war, was another population census. This was undertaken in November 1973 and, as in 1963, there were exceptional procedures intended to ensure accuracy; on this occasion, armed soldiers accompanied each enumerator. The result, declared 'provisionally' and after a delay of several months, was a national population of 79.76 million. Public attention was held less by this implausible total than by its composition. By now the regions of the federation had been replaced by twelve States. Calculating from the official figures for 1963, population in the six States corresponding to the old Northern Region had apparently grown by 5.6 per cent on annual average, and in three of those States the totals had almost doubled in the decade. In contrast, the annual average rate of growth in the six States that had replaced the Eastern and Western Regions appeared as less than 1 per cent, and in two of those States population was recorded as having fallen since 1963. The northern States were now credited with over 64 per cent of the national population. These results provoked an outcry, could not be explained, were felt to be untenable, and were finally quashed.[3] Counting the people was removed from the military government's agenda, and constitutional arrangements have since been made, and economic policies shaped, on the basis of the results declared in

1963, whatever they might be worth; an undertaking by the civilian government elected in 1979 to hold a census in 1983 was subsequently retracted.

The World Bank's estimate of 82.6 million as the Nigerian population in 1979, shown in Table 1, can be reached by applying a 2.5 per cent annual rate of increase to the declared census result of 1963. The Statistical Office of the United Nations gives a lower estimate of the Nigerian population in 1979, 74.8 million, and, mainly for this reason, arrives at a lower total for mainland West Africa as a whole – 136.64 million as against the World Bank's 144.5 million.[4]

If the Nigerian census result of 1952–3 were raised by 2 per cent annually until 1962 and by 2.5 per cent annually between 1963 and 1979 (they being the rates of increase that have been assumed by the Nigerian government planners), the 1979 total would be only 55.3 million – over 27 million less than the World Bank estimate and nearly 20 million less than the United Nations estimate. But the 1952–3 result is believed to have been a substantial under-count. The United Nations figure for 1979 could be reached, using the rates of increase cited above, if under-enumeration by 26 per cent in 1952–3 were assumed. Alternatively, a lower proportion of under-enumeration but faster rates of growth could be assumed. More extreme assumptions would obviously be required to reach the World Bank estimate for 1979.

Thus, the size of the West African population is not known even approximately, but the United Nations estimate of 136.64 million in 1979, including 74.8 million in Nigeria, seems unlikely to be an under-estimate.

The rate of increase of the population is perhaps known within about one-half of a percentage point, both for West Africa as a whole and for most of the countries in the region. The *Demographic Yearbook* of the United Nations offers an 'estimate of the order of magnitude' of 3 per cent annually for West Africa as a whole in the period 1970–5. This rate compares with 1.8 per cent for the world total and is an estimate exceeded only in Central America. For the later 1970s, the United Nations estimate for West Africa would presumably be even higher, since the *Demographic Yearbook* gives an estimate of 3.2 per cent for Nigeria in the period 1975–9, along with 3.5 per cent for Ghana and 4.2 per cent for the Ivory Coast. On the other hand, the World Bank's estimates of annual rates of increase for the whole decade of the 1970s are 2.5 per cent for

Nigeria, 3 per cent for Ghana, and 5.5 per cent for the Ivory Coast.[5] Most of the estimates offered by these and other authorities for West African countries in the 1970s range between 2.5 and 3.5 per cent. Above that range are most of the estimates for the Ivory Coast (because of immigration) and below it are some of the estimates for Upper Volta (because of emigration), Guinea, Guinea-Bissau, Mauritania and Sierra Leone.

A rapid rate of increase in West African population results from the coincidence of birth rates which are among the highest in the world and death rates which, while high relatively to all regions of the world outside tropical Africa, are now markedly below the birth rates. The United Nations order of magnitude for the West African birth rate is 49 per thousand, and that for the death rate 21. (The equivalent estimates for North America are 16 and 9 respectively, and those for Western Europe 14 and 11.) Estimates for years in the later 1970s give a birth rate below the high 40s only in Guinea-Bissau (39–41) and possibly Sierra Leone (42–48). The number of births expected per woman during the child-bearing period lies in a range between 6 and 7 for all West African countries except Guinea-Bissau; even at the lower end, this rate is over three times as great as in America and Great Britain.

More variation appears in estimates for the death rate, with Ghana, the Ivory Coast, Togo and possibly Liberia and Nigeria below 21 per thousand, and The Gambia, Guinea-Bissau, Mali, Mauritania, Niger, Upper Volta and possibly Guinea above that figure. There is a high incidence of death in the early years of life. Estimates of infant mortality rates (deaths in the first year of life per thousand live births) are in a range of roughly 100 to 200. Most of the figures come from sample surveys conducted in the late 1950s or early 1960s, but later findings include 115 in Ghana in 1970, 159 in Liberia in 1971 and 217 in The Gambia in 1973. In contrast, the infant mortality rate was only 13.0 in the United States in 1979, and 12.6 in England and Wales. This contrast is proportionately even greater in the child mortality rate (deaths per thousand after the first year of life but below the age of five). The World Bank estimates this rate lying between lows of 16 in Liberia and 22 in Ghana and Nigeria and highs of over 30 in the countries of the Sahel,[6] while in America and England the rate is well below 1.

It appears possible that one-half or more of the total mortality in West Africa occurs (or did, in the 1960s, occur) in the first five years of life, as against only 2 or 3 per cent in America and England; Table

Table 9 Mortality by age-groups

	Under 1 year (%)	*1 to 4 years (%)*	*5 years and over (%)*	*Total deaths recorded*
Benin 1961*	22.6	25.0	52.4	54,226
Ghana 1971†	17.2	19.4	63.4	30,252
Guinea-Bissau 1970	30.0	12.0	58.0	1,605
Lagos (Nigeria) 1969	33.6	22.3	44.1	6,160
Liberia 1970*	42.6	14.8	42.6	24,990
Togo 1961*	23.9	25.2	50.9	44,710
USA 1976	2.5	0.5	97.0	1,909,440
England & Wales 1977	1.4	0.2	98.4	575,928

Notes: * Results obtained from sample survey.
† Results considered unrepresentative because they are for compulsory registration centres only, serving approximately 35 per cent of the population in 1969.

Source: Calculated from data in United Nations, *Demographic Yearbook* 1979, Table 19.

9 marshals the fragmentary available information on this point.

High mortality in the early years of life implies a relatively low expectation of life at birth. Sample surveys conducted in Togo and Upper Volta at the beginning of the 1960s, and in the rural areas of Nigeria in 1965–6, put this value at between 31 and 38 years. Estimates by the United Nations Statistical Office for the early 1970s include 39 years for males and 42 years for females in The Gambia, Guinea and the Sahelian countries, and 42 and 45 years respectively in Ghana, the Ivory Coast and Sierra Leone. A population growth survey in Liberia gives for 1971, contrary to the usual experience, a higher estimate for males than for females – 46 and 44 years respectively. The World Bank has made estimates for 1979 ranging between 42 years for The Gambia and Guinea-Bissau and 54 for Liberia; the Sahelian countries (43) are near the bottom of this range, and Ghana and Nigeria (49) near the top.

Although these estimates of life expectation at birth suggest considerable improvement in West Africa since 1960, they still fall far below the figures calculated for the United States (68.7 for males and 76.5 for females in 1975) and England and Wales (69.6 and 75.8 in 1974–6).

Some figures are also available for West Africa of the expectation of *life remaining* at specified ages. They are derived from the sample surveys conducted in Benin, Togo and Upper Volta in 1960–1, the

Ghanaian census of 1960, the Nigerian rural demographic survey of 1965–6, and the Liberian population growth survey for 1971.[7] In the United States and the United Kingdom, the expectation of life remaining reaches a maximum at the age of one, when it is very slightly higher than at birth. The West African evidence, such as it is, shows considerable increases – of between 3.5 and 10.9 years – in the expectation of life between birth and the age of one. The largest increase was for males in Ghana in 1960. The latest data (Liberia in 1971) still show large increases – 6.3 years for males and 8.7 years for females. Except in Ghana, further increases (up to 6.3 years) then appear between the ages of one and five, and it is at the latter age that the expectation of life remaining reaches a maximum. This pattern of changing life expectancy in West Africa reflects, of course, the information available on age-specific mortality rates, and suffers the same limitations as to coverage and dating.

The West African populations are still predominantly rural in habitation. The United Nations estimate for the ratio of urban to total population in the region as a whole was only 18.5 per cent in 1975.[8] The censuses taken in the 1970s, and later estimates, suggest that about one-third of the populations may be urbanized in The Gambia, Ghana, the Ivory Coast, Liberia and Senegal. In Nigeria, there is little basis for making an estimate, but a proportion of about one-fifth is sometimes suggested, and Guinea, Mauritania, Sierra Leone and Togo may also be about this mark. Lower urban ratios are likely in Benin, Mali, Niger and Upper Volta; in the last, the census of 1975 showed only 6.4 per cent of the population as urbanized.

It is generally believed that the urban populations have been growing at double or more the national rates. For the region as a whole, the growth rate of urban population has been estimated at 6.9 per cent annually in 1950–60 and at 6.2 per cent in 1960–70;[9] and it is unlikely to have fallen significantly since 1970.

The consequent rise in the proportion of the population living in towns may be expected to have profound demographic and other social consequences in the long term. It is plausible to assume, for instance, that infant and child mortality are lowered by an urban environment, that the desire to procreate is less where livelihoods are gained outside agricultural self-employment, and that conceptions of familial obligation become narrowed in the cosmopolitan ambience of the city. As yet, empirical information to support these assumptions is not abundant in West Africa, though there are certainly grounds for believing that death rates are lower in the cities than in the rural areas.[10]

Big urban concentrations are relatively few in West Africa, however important they may be in the economic and political life of the countries in which they are situated. The *Demographic Yearbook* lists only some forty towns in the region having populations (according to the latest available estimates) in excess of 100,000 while in England and Wales alone there were 143 such towns in 1977.[11] Twenty-seven of the West African list are Nigerian towns, and over half of them are long-established Yoruba settlements in which town life has been compatible with agricultural pursuits for many residents. Nevertheless, there are now a few very large West African cities. Lagos is credited with a population of one million in the city proper in 1975, while the conurbation including Lagos probably contained over 1.5 million. Also in Nigeria, Ibadan at the same date is credited with 847,000 people, Ogbomosho with 432,000 and Kano with 399,000. These and other cities have been growing fast in recent years. Outside Nigeria, the 1976 Senegalese census showed a population of 799,000 in Dakar, and Abidjan in the Ivory Coast had a 1975 census population of 686,000. The 1970 census population of Accra in Ghana was 564,000, and that of the metropolitan area including Accra and the port of Tema was 738,000.

The economically active population and its dependants

A population with a high rate of natural increase is a youthful population. This is so even where the population growth results from decline in the death rate rather than from increase in the birth rate, as is the case in West Africa, because the fall in mortality occurs mainly in the early years of life. While ages are among the least reliable demographic information where births are not generally registered, national data on the age composition of West African populations are consistent in showing 40 to 50 per cent of the total population to be under the age of fifteen.[12] For West Africa as a whole in 1975, the United Nations Statistical Office estimates 46 per cent of the population to be under fifteen, 52 per cent to be aged between fifteen and sixty-four, and 3 per cent to be sixty-five and over. The corresponding estimates for North America are 25, 65 and 10 per cent respectively, and for Western Europe 23, 63 and 14 per cent; in North America, the annual rate of population growth in 1970–5 is put at only 0.9 per cent, and in Western Europe at 0.6 per cent, compared with 3.0 per cent in West Africa.

If it is assumed that people below the age of fifteen and above the age of sixty-four do not sustain themselves at all, their number, expressed as a percentage of those between these ages, becomes the so-called dependency ratio – the number of dependants supported by every hundred people of 'working age'. The estimates cited above imply a dependency ratio of 94 in West Africa, compared with 54 in North America and 59 in Western Europe. Such a high ratio as the West African is sometimes represented as a price paid for a rapidly increasing and therefore youthful population. But international comparisons may be deceptive here as elsewhere. It cannot reasonably be assumed in West Africa, as perhaps it can in richer countries, that nearly everyone is fully occupied by schooling between the ages of five or six and fourteen, and is in retirement above the age of sixty-four. It may be true that the young and the old are generally less productive than those of 'working age', but in that case the age structure is telling something about productivity rather than about the ratio of dependants. Further, whatever the true dependency ratio may be, it is not clear why West African parents, who presumably wish their children to survive, should regard themselves as better off if the ratio were smaller.

The economically active population, or labour force, is defined both more and less extensively than the population of 'working age' – more extensively because, at least in principle, it is not confined within age limits; less extensively because it excludes students and women occupied solely in domestic tasks as well as retired persons and persons wholly dependent on others. The relative size of the economically active population can be obtained from the results of censuses or sample surveys and from official estimates. In the United States and the United Kingdom, 47 to 48 per cent of the total population were reported as economically active in the late 1970s. The equivalent ratios might be expected to be lower in West Africa, because of the more youthful age structures. It is not clear that this expectation is borne out. On the one hand, there is an official estimate of 37 per cent for Nigeria in 1981. Also, the Ghana census of 1970 produced a ratio of 38.9 per cent – but on the assumption that no person under the age of fifteen was economically active. The Ivorian census of 1975 gave a figure of 42.2 per cent, but here there was an even more important exclusion from the labour force – that of all unpaid family workers. On the other hand, the official

Table 10 Structure of economically active population

A. GHANA, 1 MARCH 1970	*Wage and salary-earners*	*Other labour force*
Number	*750,000*	*2,581,000*
	percentages	
Professional, technical and related workers	3.2	0.5
Administrative and managerial workers	0.2	0.2
Clerical and related workers	2.5	0.2
Sales workers	0.7	11.8
Service workers	2.3	0.4
Agricultural, animal husbandry and forestry workers and fishermen	5.4	49.4
Non-agricultural production workers and transport operatives	8.4	10.3
Persons seeking work for the first time	n/a	4.7
	22.5	77.5

B. IVORY COAST, 1970	*Wage and salary-earners (formal)*	*Wage and salary-earners (informal)*	*Other labour force*
Number	*256,000*	*335,000*	*1,712,000*
		percentages	
Primary activities	2.4	12.6	63.2
Secondary activities	3.6	0.7	5.0
Tertiary activities	5.1	1.2	1.1
Unemployed (urban)	n/a	n/a	5.0
	11.1	14.5	74.3

C. LIBERIA, 1 FEBRUARY 1974

Number	*Total labour force 433,000*
	percentages
Agriculture, hunting, forestry and fishing	71.6
Mining and quarrying	5.1
Manufacturing	1.3
Electricity, gas and water	0.1
Construction	0.9
Trade, restaurants and hotels	3.8
Transport, storage and communications	1.5
Financing, insurance, real estate and business services	0.3
Community, social and personal services	10.0
Persons seeking work for the first time	5.4
	100.0

D. NIGERIA, 1975

Number	*Wage and salary earners (modern sector) 1,500,000*	*Other labour force 27,720,000*
	percentages	
Agriculture	0.4	60.8
Mining and quarrying	0.3	0.1
Manufacturing and processing	1.1	14.9
Electricity, gas and water	0.1	n/a
Construction and building	0.7	0.1
Distribution	0.3	11.3
Transport and communications	0.3	0.3
Services	1.9	2.9
Unemployed	n/a	4.5
	5.1	94.9

Sources: Ghana (census results) from ILO, *Year Book of Labour Statistics 1976*, Table 2B; Ivory Coast from Ministère du Plan, *L'image base 1970 – Emploi, éducation, formation*, vol. I (Paris: Société d'Etudes Economiques et Financières, 1973), cited in Heather Joshi *et al.*, *Abidjan: Urban Development and Employment in the Ivory Coast* (Geneva: ILO, 1976), Statistical Appx., Tables 24 and 26; Liberia (census results) from ILO, *Year Book of Labour Statistics 1980*, Table 2A; Nigeria from *Third National Development Plan 1975–80* (Lagos: Central Planning Office, 1975), vol. I, pp. 367–9.

estimate for Niger in 1978 is 50.8 per cent – higher than the American and British figures. Most curiously, in Upper Volta, an official estimate of 53.1 per cent in 1972 (including 56.4 per cent for the female population) was replaced by a 1975 census finding of 22.9 per cent (including 1.6 per cent for the female population).[13]

As with the statistics showing the sectoral origin of GDP (Table 3), so with the proportions of economically active population, the variations among West African countries are such as to make it improbable that real differences are being represented. The measurement, and even the conception, of an economically active population are bound to be uncertain where, for most people, livelihoods are informally organized and unrecorded, and where for many the getting of them is also intermittent, part-time, or not easily separable from activities deemed not to be economic. The number of wage and salary earners in West Africa – men and women in what might be termed regular jobs – is relatively small. About 1960, it was put at roughly two million, about 6 per cent of the labour force then estimated for the region.[14] The number has risen since then both absolutely and relatively, but not by so much as to make wage-employment the predominant means of earning an income. In Ghana, wage-employment accounted for 20 per cent of the labour force in 1960 and 22.5 per cent in 1970, according to the census results. For the Ivory Coast, this proportion has been estimated at 22 per cent in 1965 and 26 per cent in 1970.[15] Elsewhere, wage-employment is relatively less important. In Nigeria, it was estimated at 9.3 per cent of the labour force in 1981.[16] Most West Africans continue to find their livelihoods through self-employment or work in household or family enterprises which is not formally remunerated. In this context, the distinction between those who are active and those who are inactive economically is bound to be arbitrary. This arbitrariness is probably the main explanation of the disparities noticed above.

Similar uncertainties arise in attempts to determine the distribution of the economically active population among economic sectors. Table 10 presents such information on the structure of the labour force for four West African countries. The Ghanaian and Liberian figures are results of the censuses in 1970 and 1974 respectively. Those for the Ivory Coast in 1970 are official estimates of the planning ministry. The estimates for Nigeria in 1975 come from the third national plan and are based on findings of the rural demographic survey of 1965–6 and a labour force sample survey carried out in 1966–7.

In the Ghanaian, Ivorian and Nigerian figures, a distinction is made between wage- and salary-earners and the rest of the labour force comprising self-employed persons, family labour, unpaid apprentices and persons seeking work. But the Nigerian figures distinguish only wage- and salary-earners in the modern sector, or formalized employment; another 680,000 persons estimated to be working for hire in small establishments (110,000 in agriculture and 570,000 outside it) are included with the rest of the labour force. The Ivorian estimates show wage- and salary-earners in formal and informal employment separately. It will be observed that, according to these estimates, informal wage-employment was greater than the formal in the Ivory Coast (335,000 against 256,000) but less than half as great as the formal in Nigeria (680,000 against 1.5 million).

The first comparison possible is of wage-employment in the formal or modern sectors of the Ivory Coast and Nigeria. In the Ivory Coast, the proportions work out at 22, 32 and 46 per cent in primary, secondary and tertiary activities respectively, while in Nigeria they are 7, 44 and 49 per cent. The difference appears not implausible, since large-scale agricultural enterprises might be expected to have been relatively more important in the Ivory Coast in 1970 than they were in Nigeria in 1975.

Second, the distribution of all wage-employment – both formal and informal – can be compared for the Ivory Coast, Nigeria and Ghana between agriculture (or primary activities) and other activities. In the Ivory Coast, the division is 58 per cent in agriculture and 42 per cent outside. In Nigeria, taking into account the estimate of 680,000 employees in small establishments, it becomes 10 and 90 per cent. In Ghana, the division is 24 per cent in primary and 76 per cent in other activities, if it is assumed that the first three occupational categories (professional and technical, administrative and managerial, and clerical and related workers) belong entirely to the secondary and tertiary sectors. The apparent differences in the relative importance of agricultural wage-employment produce the ordering that might be expected, with the Ivory Coast first and Nigeria last, but the differences are surprisingly great.

Third, the distribution of the entire estimated labour force can be compared for all four countries. The proportion in search of work appears about the same (5 per cent) in each. In the Ivory Coast, the distribution among primary, secondary and tertiary activities works out at 78, 9 and 7 per cent respectively. In Liberia, it is 72, 7 and 16 per cent. In Nigeria, 61, 17 and 17 per cent. In Ghana, if one-third of the number in the first three occupational categories is attributed to secondary activities and two-thirds to tertiary activities, as would seem reasonable, the proportions become 54, 21 and 20 per cent.

The contrast among these distributions obviously lies in the relatively large share of agricultural occupations and the small shares of non-agricultural occupations in the Ivory Coast (and, to a less extent, Liberia) as compared with Ghana (and, to a less extent, Nigeria). This contrast is the more remarkable when it is remembered that, according to the estimates of the sectoral origin of GDP given in Tables 3 and 4 for years in the 1970s, the Ivory Coast and Liberia are countries with relatively *small* proportions of their products emanating from primary activities and relatively *large* proportions emanating from tertiary activities, while in Ghana the case is just the opposite, with agriculture making a relatively very large contribution to the GDP and services a relatively small contribution.

International differences in the productivity of labour among the sectors are unlikely to explain more than a little of these apparent inconsistencies. Probably the contrasts observed in the relative importance of agricultural and non-agricultural employments, especially as concerns workers other than wage-earners, are substantially spurious, the results of uncertainty in allocating workers among classes of economic activity in countries where industrial organization is undeveloped and occupational specialization unusual. Even the comparative data for wage-earners appear not wholly convincing, and one may suspect in particular that informal wage-employment in Nigeria in 1975 was underestimated.

At all events, it is noteworthy that in all four countries (which, it will be remembered, are among the more commercialized in West Africa) the majority of the labour force is attributed to agriculture and other primary activities. The estimates of about 55 to 80 per cent of labour in this category compare with only about 3 per cent in the United States and the United Kingdom. Even in a West Africa where mining and manufacturing have assumed importance in the estimation of GDP, it is no doubt still true that agriculture and animal husbandry and the trade in their products continue to provide most of the livelihoods of most people.

Population growth and labour productivity

Whatever uncertainties there may be about the relative size of the economically active population in West African countries, there is obviously ground for supposing that it would be greater if the rate of population growth were less and the average age of the population consequently higher. On this ground, and supposing labour productivity to remain constant, output per head of population would apparently be raised by a deceleration in population growth and lowered by an acceleration. Supposing the productivity of the labour force were rising, even rising faster than population, this inverse relationship between output per head of population and population growth would appear still to hold – output per head of population would grow faster if population grew slower. This reasoning would break down only if productivity were itself promoted by a quickening of population growth.

The relationship to which attention must be directed is therefore that between population growth and labour productivity. There are arguments showing population growth to be inimical to productivity. It used to be commonplace to assume diminishing returns from land. Assuming the supply of land to be fixed and techniques to be constant, growth in the agricultural labour force would eventually result in falling marginal productivity of labour. But in West Africa the area of land under cultivation has not been fixed but has grown over time and might be further extended, and techniques have been adapted to changing circumstances and could be further developed. There are pockets of population pressure on land in the region, not entirely relieved by migration, but West Africa as a whole has not suffered a surfeit of agricultural labour and is in no immediate danger from this indulgence; indeed, complaints are regularly expressed of scarcity of such labour in many parts of the region.

The more weight attaches, therefore, to a second argument, associated with the concept of the dependency ratio already noticed, that productivity is depressed in countries experiencing population growth by diversion of resources from productive capital formation to the nurture of dependants. The faster the rate of natural increase in the population, the greater becomes this drain into feeding and schooling the young of resources that might have been used to make labour more productive by providing more tools, machines, vehicles and buildings. Conversely, a deceleration in population growth would allow investment to be stepped up and productivity to rise faster.

Of course, this argument can undergo a *reductio ad absurdum*. Even a stationary population would have dependants toward whose nurture resources would have to be diverted. Stopping this 'loss' of resources would require an end of procreation.[17] But then the eventual result, in principle, would be a capital stock without a labour force to operate it and a population to benefit from it. In practice, no doubt, there would be little incentive to create capital, or at least the longer-lived kinds of assets, if procreation ceased. Evidently, therefore, productivity is not to be maximized by minimizing the proportion of dependants in the population. The desirability of increasing the capital stock needs to be reconciled with the desirability of maintaining, and possibly even increasing, the labour force.

It is not only at the logical extremity just mentioned that it is a mistake to take the demand for capital for granted. The argument under review does so presume. It regards resources used for the nurture of dependants as lost to productive capital formation. It further assumes that these resources would be fruitfully used if they could be secured for productive investment. Yet there has been frequent difficulty in West Africa, as will appear in Chapter 6, in finding good uses for funds available for investment, and often the uses deemed to be good have disappointed expectations. Reduction in the proportion of dependants through slower population growth would certainly not be a sufficient condition of rising productivity. In theorizing about economic growth, the effective demand for capital, or investment opportunities, need to be taken into account as well as, and perhaps rather than, the supply of investible resources.

Among the forces that may discover new investment opportunities and raise the effective demand for capital is population growth. The increase of numbers will not achieve this result unaided. But it is among the pressures to which creative economic response is possible, and indeed has often occurred. Population growth (perhaps more especially, a sudden acceleration in that growth) may be a stimulus or inducement to investment and other means of raising productivity; it has been called 'demonstrably . . . an integral part of the development process in all countries that are economically advanced today'.[18] If, on the other hand, it competes for investible resources through the rise in the proportion of dependants, this may count for less in securing gains in productivity than does the heightened demand for those resources. Possibly the historical lack of this demographic pressure is among the reasons why West Africa became economically backward.[19]

It might be thought that the question whether population growth impedes or assists economic growth could be resolved by international comparison of the growth rates in population and production. Those estimates, for whatever they may be worth, show no clear correlation between growth in population and growth in output per head. For instance, among the West African countries believed to be experiencing the fastest rates of population growth are two (the Ivory Coast and Nigeria) which, according to the estimates given in Table 7, have enjoyed rapid growth in GNP per head since 1960, and one (Ghana) which has apparently suffered decline in GNP per head. Similarly, the countries credited with rates of population growth below the West African average include both fast growers economically (Guinea-Bissau and Mauritania) and slow growers (Guinea, Sierra Leone and Upper Volta). For a longer period (1950–75) and much wider selection of countries, the World Bank estimates presented by Morawetz[20] are similarly ambiguous. Historical (as opposed to cross-sectional) data indicate no less convincingly the absence of any general association, negative or positive, between population growth and the increase in output per head.[21]

The explanation that is given for the failure of empirical evidence generally to support either of the hypotheses under consideration is that changes in the size and age-structure of the population are only one of many forces bearing on the productivity of labour. Hence, where rapid population growth is found compatible with rapid increase in output per head, it is still possible to argue that economic growth would have been even faster if the population growth had been slower. Population growth may be tending to pull down output per head, even though in practice output per head does not fall. Contrariwise, of course, population growth might be tending to push up output per head even though in practice output per head did not rise. The issue cannot be settled by estimates of population and aggregate production, and could not even if those estimates were more reliable than they are.

It might be concluded that the relationship between population growth and economic growth is not only uncertain but often rather weak – that the increase of the population is not usually a dominating influence on economic performance. This is not an orthodox view. Specifically, rapid population growth in developing countries is commonly regarded as both important for and prejudicial to economic performance. Thus, a World Bank report on African development, published in 1981, states unequivocally that: 'The consequences of rapid population growth for economic development and welfare are very negative . . . Thus, it is crucial to take steps now to reduce fertility.'[22]

The World Bank projects that West Africa might contain a population of 273 million, including 161 million Nigerians, by the year 2000, and that, twenty years beyond that date, the Nigerians alone might number 341 millions.[23] Arresting as these numbers are, they tell nothing of what happens to economic development and welfare. The precise grounds on which the World Bank report appears to predicate very negative consequences are continuing urbanization, leading to the lack of such basic amenities as water, sanitation and electricity by most urban inhabitants; increased food imports; and a slowing of progress toward universal schooling and health care. There are some puzzles in this indictment. It is not clear why the purchase abroad of more food should be regarded as a detraction from economic development and welfare. Against the postulated slowing of progress toward universal health care may be set the causes of the population growth, stated to be reduction in infant and child mortality rates by improved health and nutrition, and possibly also increased fertility because of better maternal health. But perhaps it is apparent that the negative consequences that have been adduced concern specifically public administration; they are the administrative and political problems of trying to keep provision of public services in line with the growing numbers claiming those services. A tendency to identify the problems of public administrators with economic and social retrogression is common in the discussion of population growth, and will be met with again in connection with urban in-migration. No doubt these problems are important. Rapid growth in the number of claimants is an additional difficulty that the rather frail state apparatuses of West Africa could well be spared. It may still be doubted that these problems provide sufficient ground for the belief that population growth in the region is, or will be, inimical to the increase of labour productivity and output per head.

Income, life expectancy, and schooling

Even if population growth were to detract from the increase of output per head, it is not evident that the extra numbers should be regarded as a cost suffered by the society receiving them. The kinds of goods and services summed in the GNP do not represent the only sources of human satisfaction. Increase of population through maintenance or increase of a rate of fertility in face of falling infant and child mortality rates may be a source of values far greater, to those personally concerned, than any loss sustained in economic output averaged per head.

Mortality among the young in West Africa, while still so high relatively to most other parts of the world, must have fallen greatly to have produced rates of natural increase in population now estimated at 3 per cent or more per year. Commensurately with the fall in these mortality rates, the expectation of life at birth has lengthened. In Ghana, for example, this value is estimated to have risen from 39.5 years in 1948 to 45.5 in 1960 and 49 in 1970 and later years.[24] More generally, the United Nations experts estimated life expectancy at birth to average around 30 years in West Africa in the 1950s,[25] while by 1979 most estimates for countries in the region were in a range from 40 to 50 years. Life expectancy must have been even shorter in, say, the 1930s, before population growth in the region became very perceptible. Possibly since then the expectation of life for a child born in West Africa has risen by about 80 per cent on average. As an alleviation of the human condition in West Africa, gains in this dimension surely count for at least as much as changes in the flow of economic goods and services. As between areas like West Africa and the countries rich in material assets, they manifest a narrowing gap not revealed by estimates of income per head and from which, indeed, national accounting distracts attention.[26]

Another dimension along which wellbeing may be improved is leisure. In West Africa, the possibilities do not lie mainly in reduction of working hours for the economically active population; on the contrary, the increasing commercialization of economic life could even diminish the time available as leisure to this group. The gains are rather to be won in two other ways. First, the labour time required, mainly of women and children, to perform intra-household services can be reduced by improvements in housing and

Table 11 Numbers enrolled in schools as percentages of age-groups, 1960 and *c.* 1977

	Primary schools		*Secondary schools*	
Benin	1960	1979	1960	1979
	6–11 yrs	*5–10 yrs*	*12–18 yrs*	*11–17 yrs*
Total	26	60	2	12
Males	38	78	2	18
Females	15	42	1	7
The Gambia	1960	1978	1960	1978
	6–11 yrs	*8–13 yrs*	*12–18 yrs*	*14–19 yrs*
Total	12	37	3	12
Males	17	50	4	17
Females	8	24	2	7
Ghana	1960	1977	1960	1977
	6–15 yrs	*6–11 yrs*	*16–19 yrs*	*12–19 yrs*
Total	38	71	5	32
Males	52	80	9	39
Females	25	61	2	24
Guinea	1960	1971	1960	1971
	7–10 yrs	*7–12 yrs*	*11–18 yrs*	*13–18 yrs*
Total	30	28	2	14
Males	44	39	3	21
Females	16	18	–	6
Guinea-Bissau	1960	1977	1960	1977
	6–11 yrs	*6–11 yrs*	*12–16 yrs*	*12–16 yrs*
Total	25	112	3	10
Males	35	152	3	16
Females	15	72	2	4
Ivory Coast	1960	1977	1960	1977
	6–11 yrs	*6–11 yrs*	*12–18 yrs*	*12–18 yrs*
Total	46	71	2	14
Males	68	88	4	21
Females	24	54	1	7
Liberia	1960	1978	1960	1978
	6–11 yrs	*6–11 yrs*	*12–17 yrs*	*12–17 yrs*
Total	31	64	2	20
Males	45	80	3	29
Females	18	48	1	12
Mali	1960	1977	1960	1977
	6–11 yrs	*6–11 yrs*	*12–17 yrs*	*12–17 yrs*
Total	10	28	1	9
Males	14	36	1	13
Females	6	20	–	5

Mauritania	1960	1977	1960	1977
	6–11 yrs	*6–12 yrs*	*12–18 yrs*	*13–19 yrs*
Total	8	26	–	5
Males	13	34	1	9
Females	3	17	–	1
Niger	1960	1977	1960	1977
	6–11 yrs	*7–12 yrs*	*12–18 yrs*	*13–19 yrs*
Total	5	22	–	3
Males	7	29	–	4
Females	3	16	–	2
Nigeria	1960	1977	1960	1976
	6–12 yrs	*6–12 yrs*	*13–17 yrs*	*13–17 yrs*
Total	36	69	4	13
Males	46	n/a	6	n/a
Females	27	n/a	1	n/a
Senegal	1960	1977	1960	1975
	6–11 yrs	*6–11 yrs*	*12–18 yrs*	*12–18 yrs*
Total	27	41	3	10
Males	36	50	4	14
Females	17	32	2	6
Sierra Leone	1960	1977	1960	1977
	6–12 yrs	*5–11 yrs*	*13–19 yrs*	*12–18 yrs*
Total	23	37	2	12
Males	30	45	3	16
Females	15	30	2	8
Togo	1960	1977	1960	1977
	6–11 yrs	*6–11 yrs*	*12–18 yrs*	*12–18 yrs*
Total	44	102	2	25
Males	63	129	4	39
Females	24	75	1	12
Upper Volta	1960	1978	1960	1978
	6–11 yrs	*7–12 yrs*	*12–18 yrs*	*13–19 yrs*
Total	8	17	–	2
Males	12	21	1	3
Females	5	12	–	1

Notes: – indicates less than 0.5 per cent.
n/a indicates not available.

Source: UNESCO, *Statistical Yearbooks* 1978–79 and 1980, Table 3.2.

public amenities – above all, in Africa, by better supply of water. Second, leisure can also be enhanced by deferring the age of entry to the labour force, or reducing the participation of the young in economic activity, usually done through the extension of schooling.[27]

The extension of schooling is significant on several counts. It can be regarded as a form of human capital formation – a view which accords, in fact, with the expectations of West Africans, where they have financed privately the education of their children. The creation of literacy by schooling can be viewed as the indispensable means of giving access to widespread cultures and modern forms of economic activity. Schooling can be seen as a powerful force raising aspirations and encouraging refusal to accommodate to poverty, especially rural poverty, with results that may be economically constructive, socially disruptive, or both. But it is enough for present purposes to observe that, when children are in school, they are freed from economic activity and from intra-household services.

In the period since the Second World War, there has been a great expansion in educational provision in West Africa, both in absolute numbers and relatively to population. Thus, between 1960 and about 1977, primary school enrolments in the region as a whole rose from about 4.6 million to 14.1 million. Over the same period, secondary school enrolments grew from about 200,000 to over 2 million. Teachers in training increased from some 37,000 to 159,000, and students in higher education from 6,000 to 117,000.[28] Much of this expansion has occurred in Nigeria – not only because of the size of the Nigerian relatively to the West African population, but also because of the increase in Nigerian public revenues since 1970 and a government policy of making educational provision universal at primary level. In 1960, primary school enrolments in Nigeria were about 2.9 million, secondary enrolments (including trainee teachers) 160,000, and students in higher education some 3,000. Nigerian official statistics show the corresponding figures for 1978/79 as nearly 11.5 million, 1.5 million, and 108,000, and project primary enrolments exceeding 17 million in 1984/85.[29]

Table 11 shows primary and secondary school enrolments in each West African country expressed as percentages of the relevant school-age populations in 1960 and a year about 1977. These figures are, of course, vitiated by the weakness of the demographic data, especially in Nigeria. It will be observed that they are also often not comparable, both among countries and even in the same country at the different dates, because of variation in the definition of the relevant age-group. Further, it should be noticed that these ratios are gross in the sense that they show the proportion of enrolled children *of all ages* to the estimated size of the relevant age-group. The presence in enrolments of children outside the relevant age-group explains why these ratios can exceed 100 per cent; in particular, over-age children repeating school years are numerous in the French-speaking countries. Net enrolment ratios (proportions of enrolled children of the relevant age-group to the estimated size of the relevant age-group) are available for only a few West African countries. They suggest that there may be little difference between gross and net ratios at secondary level, but that at primary level the differences can be considerable, especially where enrolment ratios are high. Thus, the gross and net ratios in 1977 (both sexes together) appear as 112 and 66 per cent respectively in Guinea-Bissau, 102 and 72 per cent in Togo, and 41 and 31 per cent in Senegal.

These qualifications notwithstanding, the figures in Table 11 are adequate to support the view that schooling has expanded rapidly in relation to population in West Africa. It might be guessed that the net primary enrolment ratio for the region as a whole roughly doubled between 1960 and 1980 – from, say, 30 to 60 per cent – while the net secondary enrolment ratio may have risen from somewhere about 3 per cent to 12 or even 15 per cent. Universal primary education, a declared objective in some parts of West Africa ever since the 1950s, could well be attained in the 1980s in Nigeria, the Ivory Coast and Togo; and in Nigeria there are plans to achieve a 40 per cent transition rate from primary to secondary schools early in the 1980s.

School enrolments are an imperfect measurement of a net increase in leisure. A child enrolled at the beginning of a school year does not necessarily attend throughout the year. Children may have to do in out-of-school hours and during holidays some of the work they have avoided through schooling. Work that would have been done by children attending school may have to be done instead by women and by children who are not sent to school – it is noteworthy that the enrolment ratios in Table 11 are often only one-half for girls of what they are for boys. Children may be not so much released from economic activity and intra-household services by rising income, as diverted into schooling at the expense of other people's time.

Even so, the time spent by pupils in school is arguably a life-enhancing use of resources that deserves to be reckoned among criteria of social amelioration. As with life expectancy, its use for this purpose produces a tendency for international disparities to narrow over time rather than to widen as is the case with estimates of GNP per head.[30]

In the long term, it might be expected that a society progressing materially would enjoy a rising income per head, would increase its number (mainly through reductions in infant and child mortality), and would use more of its time as leisure, at least so far as leisure is increased by prolongation of dependency, or increasing deferment of entry to the labour force, through the extension of schooling. It has already been shown that gains in all three of these dimensions have, in fact, been made in West Africa considered as a whole and in many countries of the region considered separately.

The freedom to choose among these forms of amelioration, and hence the power to maintain among them whatever might be regarded as an appropriate balance, may be quite limited in practice. Thus, it may be practically easier to reduce mortality rates and to extend schooling than to increase the flow of economic goods and services. It would then be possible for population to grow and leisure to increase apparently at the expense of economic output which, averaged over population, remains stationary or even falls. Ghana in the 1960s and 1970s is a case in point. The population increased by about two-thirds (from 6.7 to 11.3 million), the proportion of the school-age population in school probably more than doubled, and GNP per head fell quite perceptibly, if we are to believe the estimates (see Table 7). Perhaps more commonly, income per head grows, but at what is deemed an inadequate rate, and it is tempting to hypothesize that the economic gains would be greater if leisure were enjoyed less or if population grew more slowly. But it is by no means clear that average income could be raised, even in the long term, by slowing the rate of population growth or by increasing the participation of children in economic activity. It does appear that population can be traded against schooling; a slower rate of population growth would, as the World Bank has argued, assist progress toward universal education. Otherwise, the functional relationships among economic growth, demographic variables and schooling are not only of uncertain sign but also perhaps rather weak.

What does seem clear is that only from a narrow, or even a

distorted, viewpoint does increase in the estimate of GNP per head represent the whole of human amelioration.

Rural-urban migration

The growth in population has not, in fact, been a matter of acute political concern in West Africa. In contrast, another leading demographic feature of the region – the migration from rural areas to towns – has been a source of official anxiety ever since the 1940s.

The geographical mobility of West Africans was mentioned in Chapter 1. There is long historical evidence of movements of people to exploit natural resources better. Seasonal or temporary migration of labour, especially between the savanna and forest areas, and often crossing political boundaries, was conspicuous in the colonial period and still continues. In addition, people have moved from the countryside to permanent or long-term settlement in the towns, and especially in the largest towns. This last movement has grown since the 1940s both absolutely and relatively to other kinds of migration, and is the explanation why urban populations are increasing so much faster than national totals. It seems likely that urban population growth has been about as much the result of in-migration as of natural increase in most West African countries in the recent past; further, this migration is heavily concentrated in the principal city of each country and its neighbourhood – Accra in Ghana, Abidjan in the Ivory Coast, Freetown in Sierra Leone, Dakar in Senegal, Lomé in Togo, and so on.[31]

While migrations within the rural areas, including the seasonal or temporary movements of agricultural labour, are generally assumed to be in harmony with economic opportunities, the migration to the towns is often postulated to be excessive and out of phase with changing economic conditions, leading to unemployment and underemployment in the towns and possibly to coincident shortages of labour in rural areas. Further, the tendency for people to congregate increasingly in large towns is held, like rapid population growth, to result in diverting investible resources into social overheads at the expense of directly productive investments; relatively more has to be spent on, for example, town planning, sanitation, schooling and policing.

The migration to the towns is nevertheless judged to be economically motivated, both by most students of the subject and by the migrants themselves.[32] An urban differential has usually been postulated as the force attracting population to the towns; earnings, somehow defined and measured, have been held to be significantly higher in the towns, or in the larger towns, than elsewhere. Thus, the minimum wage paid in formalized urban employment might be shown to exceed a measurement of average rural earnings derived from household expenditure surveys;[33] or a growing differential might be inferred from comparison of indices of urban real wages and GDP per head,[34] or from the observation that, while the minimum wage had been rising, an important range of rural remuneration had been depressed by the restrictive pricing policies of marketing monopolies in export crops.[35]

Such comparisons are not enough to verify an hypothesis that migration to the towns occurs to make gains in income. More careful consideration is required both of the opportunity costs of migration and of the migrant's urban livelihood.

First, then, the migrant's supply price will vary with differences in productivity among rural areas and with distance from his urban destination. It will be lower, the less productive is the economic activity available to him in his home area and the less are his costs in reaching his destination.[36] This supply price may also differ according to whether the migrant is expecting to move permanently (and possibly having to take dependants with him) or only temporarily, and according to his age and status in his rural household (since his share in its income will not necessarily correspond with his contribution to its output). The supply price may be lowered by intra-household subsidization of the migrant, and consequently young men and women may migrate more readily from better-off than from poorer households in the same area. These considerations are enough to show that no single measurement of rural income can be obtained as a basis for explaining migration to the towns.[37]

Difficulties also arise in ascertaining the urban income relevant to migration. One difficulty is that many migrants move not simply from a village to a town, but in a series of two or more steps, going first to a nearby town, which may be small, and later to one or more of the more distant and larger towns. Small-town income may therefore be more relevant to the original decision to leave a village, but big-city income more relevant to the urban in-migration that excites most concern.

This complication could be overlooked at one time because the minimum wage legally stipulated or conventionally observed in formalized employment was taken to be the relevant urban income, and this wage was uniform among towns, or varied only according to an official scale. In practice, places in formalized employment are not easily found. It is likely that the number of workers without formalized employment in West African towns is at least as great as those with it. It is now recognized that most newcomers to the towns are absorbed initially by the so-called informal sector, finding employment as street vendors, apprentices, and the employees of small-scale manufacturing and service enterprises.

Earnings in the informal sector can be ascertained much less easily than those in formalized wage-employment. It might be expected that they would show a wide dispersion, and that they would overlap considerably with formalized wage-incomes.[38] From the nature of the informal sector, it might be expected that many of those working in it have more than one source of income. Further, there is ground for believing that many persons in formalized employment are engaged also in informal activities – a regular job being valued primarily for the regularity of the income it provides, and regarded as a complement to, rather than a substitute for, other activities.[39] Inflation since the early 1970s may be assumed to have increased the pressure on urban workers thus to diversify their sources of income. It would be hazardous, therefore, to make any general statement about the difference between wage-earnings and the incomes of other urban workers. Presumably, newcomers to a town tend to gravitate toward the least eligible employments and the lowest ranges of urban income, but how they stand in relation to the minimum wage, either in achievement or in aspiration, is much less straightforward than it appeared to analysts of the rural–urban migration in the 1950s and 1960s.

Whatever the relevant urban income might be, it would have to be adjusted in several ways if it was to be used in an economic explanation of the migration to the towns, or to a particular town. Fringe benefits supplement the wage or salary in formalized employment. They are likely to be of increasing relative importance as the ladder of remuneration is ascended, but may be of appreciable value even at the lowest rungs in such forms as paid holidays, sick leave, or subsidized transport, housing or food.[40] In the informal sector, such benefits are likely to be insignificant, but all urban residents may be expected to share in what has been called the implicit wage supplement[41] – the greater availability in the towns (or in the large towns) of free or subsidized public services such as schooling, health care, electricity, water and sanitation. For some migrants, these provisions, and more especially the better educational opportunities, may be a sufficient reason to migrate.

As an offset to these advantages, the cost of urban living (or of big-city living) may be expected to be higher. To measure the excess is difficult since the additional costs mostly reflect a different style of living and are therefore not easily distinguishable from benefits. This is true not only of outlays on goods and services unobtainable in a migrant's home area, such as electricity and commercialized entertainments, but even to some extent of those expenditures he would not need to make at home on living accommodation and food. These extra costs are not to be dismissed as irrelevant, but their proper treatment in comparing urban and rural (or big-city and small-town) incomes is not straightforward; perhaps all one can say with confidence is that they will detract less from income for a person who wishes to commit himself to a new life than for one working in the town temporarily.

Finally, the relevant urban income would have to be corrected for risk, or weighted by the probability that it will be obtained. Recognition by the migrants themselves of the need for this adjustment can be counted as one reason for their moving often in steps. The adjustment could be made most readily if migrants remained unemployed, i.e. without earnings of their own, so long as they did not obtain the work to which they aspired; the probability, or in other words the expected urban income, would then be less, the greater the number of the unemployed. But many migrants cannot afford to be unemployed. If they cannot find the work they want, they must do something less eligible. For them, whose risk is that the livelihoods they obtain in one way or another will be poorer than they hoped for, there is no similar basis for assessing probability, even though the deterrent to migration posed by risk is no less real.

In practice, this deterrent seems to be heavily discounted by many urban newcomers through the possibility of obtaining material support from kinsmen or friends already established in the town. It seems safe to say that the flow of migrants would be far less than it is, but for the willingness of townspeople to provide hospitality to those who remain unemployed or who cannot find adequate remuneration, especially the younger migrants who have travelled alone. Even if a young migrant makes no economic gain in coming to a town, he also loses nothing, provided he has kin who will support him at about the standard of living he is accustomed to and that he retains the freedom to return to his place in the household he has left.

Much informal redistribution of income occurs in consequence of the commonly acknowledged obligation of better-off persons in West Africa to assist their more indigent relatives and friends. A sample of 188 industrial wage-earners in Dakar in 1965, biased somewhat in the direction of senior employees, was found regularly to maintain a total of 1,802 persons, of whom about 1,000 comprised nuclear families but the other 800 were more distant connections. Instances of households numbering over twenty were not uncommon in this sample. While the 188 wage-earners enjoyed relatively high incomes on average, income per head in their households was found to be no more than an estimate of income per head in the groundnut areas of rural Senegal.[42] In Abidjan in 1963, a socioeconomic survey revealed a positive association between sizes of households and the earnings of household heads. Part of the explanation, no doubt, is that older men tend both to earn more and to have more wives and children. But 23 per cent of the survey population were found to be dependent members of households but neither wives nor descendants of the household heads, and it is probable that the higher income-earners attract more of this kind of dependant too. Household heads were ranked by average earnings in occupational categories descending from top administrators and managers to unskilled labourers and apprentices. While household income averaged over four times as much in the top category as in the bottom, household income per head was only some 70 per cent greater at the top than at the bottom.[43] This evidence relates only to redistribution through enlargement of urban households; in addition, income may be transferred by occasional hospitality, cash subventions and remittances.

The discussion so far has shown the difficulty of positing an urban differential in earnings as the force inducing migration to the towns when no single measurements are available either of the opportunity cost to migrants of leaving their home areas or of their expected urban earnings. The matter is further confused by the common practice of migrating in steps, only the first of which is a rural–urban movement; by the conceptual difficulty of adjusting alternative incomes for differences in living costs; by the migrant's expectation that he will benefit from subsidies in town, both through public services and in material support provided by kinsmen and friends; and by the absence of a basis (other than unemployment, which is of limited relevance) for estimating the risk that the differential will not be secured in practice. Clearly, it would be unreasonable to speak (as used to be the practice) of *the* urban differential in the sense of a general and measurable difference between urban and rural earnings.

The individual migrant knows what he is giving up and what he hopes to obtain, and judges for himself his chances of success. It might therefore be argued that an urban differential has meaning at the level of individual decision-making even if it cannot be generally observed and measured. But to represent the migrant as arriving at a decision to move through comparison of the present values of alternative streams of income, adjusted for differences in living costs and risks, seems an excessive attribution of calculating rationality to behaviour more probably resulting mainly from the sap of youth. Rationality may be overlaid by a kind of 'urban fetishism', perhaps favouring particularly movement to a capital city.[44] Glamour and excitement may count for more than estimation of gains in inducing movement to West African cities, as they are said to do in governing the supply of labour to the profession of acting. Even the rational basis of less personal investment decisions has been questioned; according to Keynes: 'most, probably, of our decisions to do something positive, the full consequences of which will be drawn out over many days to come, can only be taken as a result of animal spirits – of a spontaneous urge to action rather than inaction, and not as the outcome of a weighted average of quantitative benefits multiplied by quantitative probabilities.'[45]

This urge to action has been a powerful propellant of migration to West African towns since the 1940s, as horizons have been widened by the extension of schooling, opportunities have expanded through economic growth and political changes, and social norms have shifted in favour of urban styles of living. The large towns have become magnets for the young, many of whom may not even need to make a conscious decision to migrate, but merely find themselves staying on after one of many visits to urban relatives.

It could still be argued that a rise in wages in formalized urban employment would, other things remaining the same, induce more migration. Since the urban differential was originally postulated to have arisen through the excessive indulgence of West African governments toward wage-earners, the point is of some importance.

Not only would the towns appear still more attractive if wages were higher, but also (assuming the demand for labour in formalized employment to be inelastic with respect to wages) the capacity of the wage-earning community to carry dependants would be enlarged. This is true, but just such a stimulus to migration could also be produced by an increase in income-earning opportunities, both formal and informal, with urban wages constant or even (in real terms) falling.[46] An increase in income-earning opportunities could result from the direct and secondary effects of, for example, greater government spending in the towns. The risk of migrants' hopes being disappointed would then be reduced; the readiness of urban households to accept additional dependants would presumably be enhanced by the better chances of passing them into self-supporting activities; and the urban attraction consisting in free or subsidized public services might also be greater.

This latter combination of circumstances – increasing urban attraction in spite of constant or falling real wages – is more than a theoretical possibility. Although the real value of wages trended upward in West Africa in the immediate aftermath of the Second World War, from about 1960 they became less buoyant in money terms and underwent long periods of erosion, sometimes pronounced, in real terms, as will be shown in the next section of this chapter. It is not apparent that this change was accompanied by a decline in rural–urban migration. The elasticity of the migration could well be greater with respect to aggregate urban demand than with respect to wages in formalized employment.

In conclusion, it may be suggested that, with freedom of movement to the towns and free competition for work in informal activities, a general excess of urban over rural earnings is improbable. Indeed, if there is widespread preference on other grounds for urban life, the low end of urban earnings might be expected to compare unfavourably with many rural livelihoods. In formalized urban employment, the competition for work has been less effective in determining remuneration, since money-wages have been institutionally determined and are inflexible downward. In times when there were substantial increases in these wages and the value of money was fairly stable, it was tempting to assert that here was the signal inducing migration, moderated only by the risk that formalized employment would not in practice be obtained. But since 1960, West African governments have restrained wage increases and the real value of wages has tended to fall in face of more or less pronounced inflation, and it is now generally accepted that most newcomers to towns are absorbed by informal economic activities. Hence it is no longer plausible to represent wages at the minimum as a premium form of income attracting people to towns. Finally, the readiness of established urban residents to accept claims made on them by newly arrived kin or friends is another equalizing force. Even where differences in earnings cannot be equilibriated away by competition for jobs, it seems that differences in living standards may be much reduced by familial and fraternal obligations.[47]

No doubt migrants do move to the towns, and among towns, in the hope of bettering themselves, but this hope is likely to rest less on objective comparison of streams of expected income than on Keynes's 'spontaneous urge to action', an urge accented by new opportunities arising from economic and political changes, by the widening of horizons and stimulation of ambition through schooling, and by the incorporation into the propaganda of political parties and governments of the ethos of modernization. A migrant stands to lose little if he moves with the support of his household of origin and if an urban household is willing to receive him. His hopes of gain are likely to depend less on a perception, whether accurate or not, of a general excess of urban over rural earnings than on two other considerations: first, that it is in the towns, and especially the largest towns, that modernity is most accessible in such forms as educational opportunities, public utilities, novel recreations, and the range of goods and services available for purchase; and, second, that it is in the towns that the ladder of material advancement reaches highest, however small his own chances of ascending it may be.

Thus the problem of the urban in-migration is not a problem to the migrant, who may be assumed to have left his rural livelihood of his own volition or with the encouragement of his family, to be seeking in town to fulfil personal or familial aspirations, and to retain the freedom to return home if his hopes are wrecked. As to the cost of labour moving from agriculture where it may be scarce to the towns where it may be redundant, this is for the most part borne privately by the sending and receiving households, since there are seldom public subventions for urban unemployment and underemployment; if leisure is inadvertently increased at the expense of output, the burden is primarily on those responsible.

The problem that remains is a political and administrative

problem. Growing concentration of population may mean a more insistent political demand for urban infrastructure and social services. Disappointed hopes are less easily disregarded in the capital city than in the bush. Official decisions on the spatial allocation of resources are harder to make when the geographical distribution of the population is changing. Governments have usually had an interest in people staying where they have been put, or in their moving only at official behest; but this interest ought not to be confused with the welfare of the people or the chances of economic progress.

Wage structures and movements

The relatively small number of wage- and salary-earners in West Africa has already been remarked. They probably represent no more than one-tenth of the economically active population of the region as a whole. Many of them work for wages in the urban informal sector and in agriculture, alongside owner-operators, family labour and apprentices. The remainder, the employees of governments, public corporations and the larger private enterprises, whose terms of employment are amenable to law, government policy and possibly trade union organization, may constitute only some 5 per cent of the economically active population of West Africa, though they would be relatively more numerous in some countries of the region, including the Ivory Coast and Ghana.

The remuneration of these employees in the so-called 'modern sector' has an importance out of proportion with their numbers. Its structure is an aspect of that ladder of material advancement, the perception of which has been suggested to underlie urban in-migration. This structure is also revealing of marked differences in the values of different grades of labour in West Africa, particularly of different educational attainments. It probably makes an important contribution to the inequality of incomes, as was suggested in Chapter 2. Changes both in wages in general and in their structure, more especially in the public sector, occur as reactions to powerful political forces and important political conflicts; in Nigeria, wages and salaries have been called the most sensitive of incomes from a policy standpoint,[48] and the same could be said of other West African countries.

The differentials paid for superior grades of labour in West Africa have long been perceived to be relatively large, or in other words the range of wages and salaries is extensive. The fundamental explanation, of course, has been the scarcity in the region of educated and industrially skilled manpower. More particularly, the bulk of the administrative, managerial, professional, technically skilled and supervisory employees, in both public and private employments, has been (and in some countries still is) European, paid at rates based on what it could earn at home with additions to compensate for its displacement. In contrast, unskilled labour has been exclusively African, and its pay has been based on local opportunity costs.

There would appear to have been a powerful inducement in this wage structure to substitute African for European personnel, as the former became trained or more skilled or better adapted to the requirements of colonial administration and modern forms of industrial organization, but this substitution was retarded not only by shortage of appropriately qualified Africans, but also by doubts that their productivity really matched that of Europeans – doubts reinforced, no doubt, by prejudice.

Even nowadays, the concentration of Europeans and other non-Africans in the more senior and highly remunerated employments is reflected by large differences between their average earnings and those of Africans in formalized employments. Thus, in the Ivory Coast in 1974, non-Africans in private employment were apparently paid over six times as much as Africans on average in the tertiary sector, nearly twelve times as much in the secondary sector, and over thirty times as much in the primary sector.[49] In Liberia in the same year, non-Africans engaged in mining were recorded as receiving average earnings three times as great as salaried Liberians and four times as great as Liberian wage-earners; in construction, these differentials were much higher, about six and twelvefold respectively, and in agriculture (mainly the rubber plantations) they were higher still, more than seven and twentyfold.[50] In Nigeria, non-Nigerian managerial and professional employees in the industrial sector earned on average over twice as much as Nigerian employees in the same category, according to the *Industrial Surveys* carried out by the Federal Office of Statistics in the period 1963–74.[51]

Africanization of the better-paid jobs accelerated after the Second World War, as the local supply of suitably qualified persons slowly improved and it became politically imperative, especially in British West Africa, that use should be made of it. The impact of this process on the structure of pay was greatly restrained by a common understanding that African administrators, managers and professional employees should be accorded the same salary scales, and other conditions of service, as their European colleagues. To offer inferior terms was to be racially discriminatory. Either of their own volition or through pressure from labour organizations, employers came to accept that Africans and Europeans with equivalent responsibilities must be treated equally. In this way, it has often been said, a pay structure with differentials appropriate to a colonial order, or to a stage in economic evolution when high-level manpower had mostly to be imported, was bequeathed to independent West African countries in which economic conditions no longer justified these differentials. Substitution of African for European personnel then depended more on political than on economic pressures, and proceeded faster and further in, for example, Ghana and Nigeria than in Senegal and the Ivory Coast.

A feature of the formalized labour market in West Africa, as in other developing areas, has been the importance attached to educational qualifications. Eligibility for a job has depended less on a man's ability, proven or anticipated, satisfactorily to do the job than on whether he has successfully completed the course of schooling or training deemed appropriate to it. Several reasons can be given. Literacy in English or French was essential in many jobs. Much employment was publicly funded, and a need was felt to use objective criteria in recruiting to it. Certificates, diplomas and degrees were useful in screening applications and moderating the competition for jobs, including the competition of Africans for jobs in which Europeans predominated.

A correspondence consequently developed between the elongated pay structures of West Africa and differences in the average earnings of groups of workers at successive stages of educational attainment. It was calculated that, on average, the discounted lifetime earnings of workers who had completed primary schooling in Ghana and Nigeria in the 1960s could be expected to be more than double those of workers who had not, that four years of secondary schooling would increase this difference to a factor of four or five, and that university graduates could expect lifetime earnings more than twentyfold those of the uneducated.[52] According to a household survey conducted in Nigeria in 1967, the average earnings of workers who had completed primary education were 1.7 times those of illiterates, those of workers with secondary education were 2.7 times as much, and those of university graduates about twelve times as much.[53] These educational differentials diminished in Nigeria after the 1960s, but the premium earned by a university degree or other form of higher education remained abnormally high by international standards.[54]

In contrast, data on earnings in the private and parastatal sectors in the Ivory Coast in 1971 have been used to calculate very high internal rates of return on the secondary schooling and post-secondary training required of supervisors and technicians, but relatively modest rates of return on the university education expected of managers.[55] The explanation is that political pressure for the Ivorianization of posts was concentrated at that time in those middle-level employments, while managerial appointments were still filled predominantly by expatriates. The scarcity of Ivorian supervisors and technicians, aggravated by the rapid economic growth of the country, allowed them to command high differentials over lower grades of labour, while the willingness of the Ivorian authorities to continue to rely on high-level manpower from abroad held down the earnings of Ivorian university graduates; the earnings of a new graduate employed in management were estimated to be only 14 per cent more than those of a technician.[56] As pressure for the displacement of Europeans moves up the occupational ladder, the differentials enjoyed by Ivorian graduates, and hence the returns on university education, may be expected to increase.

This contrast between Nigeria and the Ivory Coast suggests the survival of colonial norms to be an inadequate explanation of West African wage structures. While the employment of Europeans in the colonial period played a part in creating those structures, it seems that retention of those expatriates, where they are retained, may permit a less extensive upper range of salaries than would otherwise rule. It may be added that the willingness of the Ivorian authorities also to tolerate immigration of unskilled labour has kept down wages at the lower extremity; the average earnings of non-Ivorian Africans appear consistently below those of Ivorians, especially in agricultural employment.[57]

The pay structures have therefore not been immune from market forces. They have adjusted to changes in the relative scarcity or abundance of particular grades of labour – changing more freely, it may be supposed, in private employments (and in parastatals, where they are detached from civil service pay scales) than in the public services.

In Nigeria, in particular, the pay structures have also been responsive to institutional forces, whose effect has been to compress differentials – even though they remain large by international standards. Table 12 shows for selected years from 1954 the entry salaries or wages of nine occupational titles in the Federal public service, expressed as percentages of the minimum wage for unskilled labour. It will be seen that the differentials among these

Table 12 Pay differentials in Nigeria, selected years 1954–75

A. FEDERAL PUBLIC SERVICE*					
	1954	*1965*	*1972*	*1975*	*Percentage increase in pay, 1954–75*
General labourer	100	100	100	100	393
Artisan	160	143	130	125	285
Clerical officer	205	182	165	162	288
Executive officer	773	594	475	347	121
Technical officer	814	628	500	363	120
Administrative officer	855	661	538	453	162
Engineer	1,159	892	700	495	110
Medical officer	1,216	1,024	831	607	145
Permanent Secretary	3,616	2,697	2,077	1,763	140

B. INDUSTRIAL SECTOR†					
	1963	*1965*	*1970*	*1974*	*Percentage increase in pay, 1963–74*
Manual operatives (skilled and unskilled)	100	100	100	100	134
Clerical	192	178	186	160	95
Professional/managerial: Nigerian	697	639	683	545	83
Professional/managerial: non-Nigerian	1,805	1,571	1,836	1,306	70

Notes: * The differentials are the entry wage or salary of each occupational title, expressed as a percentage of the minimum wage for general (or unskilled) labour.
† The differentials are the average wage or salary of each category shown by the *Industrial Surveys*, expressed as a percentage of the average wage of manual operatives.

Source: Adapted from Olufemi Fajana, 'Income Distribution in the Nigerian Urban Sector', in *The Political Economy of Income Distribution in Nigeria* (New York: Holmes & Meier, 1981), ed. Henry Bienen and V. P. Diejomaoh, Table 6.4, p. 208, and Table 6.7, p. 215.

occupational titles have in nearly all cases fallen with each successive year shown, and that by 1975 the differentials obtained by the higher titles relatively to the unskilled labourer's wages were roughly only one-half as great as they had been in 1954 – for example, the technical officer's pay was less than fourfold that of the general labourer instead of eightfold, and the medical officer's pay sixfold instead of twelvefold. Making the comparison in another way, the pay of the general labourer increased almost fivefold between 1954 and 1975, but that of the artisan and clerical officer increased less than fourfold and that of the higher titles by two-and-a-half times or less. This tendency toward reduction of pay differentials in the public services has continued since 1975.

The movement of differentials in formalized private employment can be less easily traced, but statistics of average industrial earnings, collected through the Industrial Surveys of the Federal Office of Statistics since 1963, show the pay of manual operatives improving relatively to that of clerical, professional and managerial employees, though the trend has been less marked and more irregular than in the public services (see Table 12B). Thus, the average earnings of Nigerian professional and managerial staff were about seven times those of manual operatives in 1963, and again in 1970, but only five-and-a-half times as great in 1974.

Wages and salaries are not, of course, the entire remuneration obtained from formalized employment. Fringe benefits are paid in such forms as the use of cars and houses, or allowances for transport and housing, and it is generally accepted that such benefits are heavily skewed in favour of the higher salary-earners. In private employment, it is possible that the tendency for pay differentials to contract has been partly offset by a widening of differentials in fringe benefits.[58] This seems unlikely to have happened in the public services, where, indeed, some attempt was made at the beginning of the 1980s to reduce the skew in distribution of these benefits.

The compression of the differentials in pay appears not to be attributable to changing economic conditions. While the number of Nigerians who have undergone higher education and technical training has grown greatly since the 1950s, this supply appears still to fall far short of demand. Thus, high-level manpower in private employment is still imported despite official discouragement, and its number has increased despite official controls. In the 1980s, as ever since the 1940s, government plans of development are acknowledged to be constrained by 'executive capacity', or shortages of administrative, professional and technical personnel capable of executing the plans.[59] In 1976 and 1977, vacancy rates for professional and technical staff (accountants, architects, engineers, statisticians, surveyors, stenographers) in the establishments of Federal ministries and public corporations were reported as mostly in a range from 45 to 70 per cent; even in the private sector,which was held to be relatively well staffed, these vacancy rates were between 20 and 30 per cent.[60]

On the other hand, the supply of unskilled labour to formalized employment is usually held to exceed demand at the going wage, so that many young men and women, especially primary school-leavers, are absorbed by the informal sector, or declare themselves to be unemployed.

The shrinking differentials in Nigeria therefore seem not to be the result of any shift in the availability of labour, relatively to employment opportunities, as between the educated or skilled and the uneducated or unskilled. Though skills remain very scarce, they have become relatively cheaper, and though unskilled labour continues to be in excess supply, it has become relatively dearer.

The explanation of this paradox is that in Nigeria, as in other West African countries, general adjustments in the level and structure of wages and salaries result from government decisions which, while not entirely detached from market forces, sometimes run counter to them. Minimum wages may be laid down, and changed, by law;[61] and, even in the absence of a legal minimum, wages have responded to official decisions because of the predominance of the public sector in formalized employment (accounting for at least one-half of the total in most West African countries) and the observance of government minimum rates by the larger private employers.

In Nigeria, the general adjustments have followed the reports of official wages commissions, which have been appointed from time to time out of recognition that the real value of wages had been much eroded by rising prices. These wage enquiries have included the Morgan Commission of 1963–4, the Adebo Commission of 1970–1, and the Udoji Commission, which recommended increases that were made retroactive to 1974. For two reasons, the reports of the commissions have led to wage increases inversely related to salaries, or in other words to a reduction in differentials. First, the trade unions and their central organizations, which have been instrumental in appointing the commissions, have explicitly and consistently urged the narrowing of pay differentials.[62] Second, the commissioners have favoured considerations of 'need' and 'social justice', rather than economic criteria, in making their recommendations,[63] having been strongly moved by the plight of the lowest-paid employees; thus, the Morgan Commission reported that workers at the lowest wages were 'living under conditions of penury', and the Adebo Commission found 'intolerable suffering at the bottom of the income scale'.[64]

The years 1965, 1972 and 1975 shown in Table 12A are those immediately following wage awards resulting from a wages commission, and the figures in the table therefore bring out the discontinuous changes in pay differentials in the public services that have been produced by these commissions. In private employments, wages have been more flexible, but general adjustments, and an associated reduction in differentials, have nevertheless tended to follow the official wage commissions because of trade union pressure for the wage awards to be enjoyed outside as well as within the public services; in 1975, for instance, every group of workers wanted its 'Udoji', regardless of increases already received in the immediately preceding years. It will be observed that the industrial wage differentials shown in Table 12B narrow in 1965 and 1974, but that in 1970 they had been restored to about the levels of 1963. The years from 1966 to 1970, which were the early period of military rule and included the civil war, were a time in which no general adjustment of wages occurred, trade union activity was severely repressed, and relative wages and salaries in private employment were particularly amenable to market forces.

The reduction in pay differentials continued in Nigeria after 1975.[65] Pay guidelines issued by the government in and after 1977 envisaged larger permissible increases for lower-paid than for higher-paid employees. The minimum remuneration of government employees was raised by nearly 40 per cent in 1980, and minimum wage legislation in 1981 raised it by another 25 per cent, and the increases in higher wages were less than in proportion.

It is to be expected that the Nigerian experience in this regard has been paralleled in other West African countries. Everywhere in the region, the government is the arbiter of wages over wide fields of employment; its judgements are strongly affected by the cost of living; and the effects of rises in the cost of living are felt more keenly, the lower is the level of remuneration. In addition, the very large pay differentials have antagonized trade unions, and they are politically difficult to defend.

While institutional forces may have been compressing pay differentials in formalized employment, they appear not to have been raising the lower end of real remuneration in most countries. The estimates given in Table 13 show the movement of minimum wages deflated by the most relevant available index of consumer prices. They mostly indicate a decline in minimum real wages between a year about 1963 and 1975 or 1980; in no case can there be said to be clearly an increase.

The figures in this table are of more than usual unreliability. There are obvious practical difficulties in ascertaining representative prices for foodstuffs and some other local products that bulk

Table 13 Evolution of real wages, *c*. 1963–80

		Index numbers	
	Base year	*1975*	*1980*
Benin	1963	66	
Ghana	1963	75 (June)	
Guinea	1959	25	
Ivory Coast (non-agricultural)	1963	114	100
Ivory Coast (agricultural)	1960	61	
Mali	1959	59	
Niger	1963	98	102
Nigeria	1973	158 (March)	95 (April 1979)
Senegal	1961	104	93
Togo	1963	99	75
Upper Volta	1960	80	92

Note: see text for explanation of data used in computing the indices.

Sources: 1975 indices for Benin, Guinea, Ivory Coast (non-agricultural), Mali, Niger, Senegal, Togo and Upper Volta from George R. Martens, 'Industrial Relations and Trade Unionism in French-Speaking West Africa', in *Industrial Relations in Africa* (London: Macmillan, 1979), ed. Ukandi G. Damachi, H. Dieter Seibel and Lester Trachtman, Table 2.13, p. 65; for Ghana from Jon Kraus, 'The Political Economy of Industrial Relations in Ghana', in ibid., Table 4.7, p. 145; and for Ivory Coast (agricultural) from Eddy Lee, 'Export-Led Rural Development: the Ivory Coast', *Development and Change* 11 (1980), Table 5, p. 626. Nigerian indices from *First Things First: Meeting the Basic Needs of the People of Nigeria* (Addis Ababa: International Labour Office, 1981), Table 49, p. 224. Other 1980 indices calculated from data in 1979 and 1980 annual reports of the *Banque centrale des états de l'Afrique de l'ouest*, Annex pp. 25–7, and linked to Martens' series.

large in workers' budgets, and the weighting of items in those budgets may not retain validity for long. In addition, only in Ghana and Nigeria has a national consumer price index been used to deflate the minimum money-wage. In the Ivory Coast (both indices), Mali, Niger, Senegal, Togo and Upper Volta, the best available price index is for African consumers in the capital city of the country concerned. In two other countries, the procedure used for estimating real wages is even more questionable. In the absence of a consumer price index for the country concerned, the real wage estimate for Benin has been obtained by averaging the price indices of Abidjan, Niamey, Dakar, Lomé and Ouagadougou, and that for Guinea is based on the price indices of Ghana and Mali on the ground that those countries also have had inconvertible currencies.

The figures must therefore be said to point toward, rather than establish, the view that minimum real wages have been trending down, or at least not trending up, in most West African countries since the early 1960s – with possibly pronounced falls by 1975 in Benin, Ghana, Guinea and Mali and among agricultural wage-earners in the Ivory Coast. (In Ghana, the fall evidently continued after 1975, since although the minimum money-wage was increased sixfold between 1975 and the end of 1980, the consumer price index rose by a factor over twice as great between those years.)

An explanation of such a trend would be the excess supply of labour generally believed to obtain in the lower grades of formalized employment. While money-wages are inflexible downward, market forces make themselves felt in a tendency for real wages to fall. Note in this connection the deterioration in agricultural relatively to non-agricultural minimum real wages in the Ivory Coast, shown in Table 13; the supply prices of wage-labour may be assumed to have been lower in the primary sector, where African immigrants predominated in the 1970s, than in other sectors where the majority of employees were nationals.

In most West African countries, the trade unions, while they may have had success in reducing pay differentials, where their objectives were congruent with public policy, have lacked sufficient political or organizational strength to raise, or even to prevent a fall in, the general level of real wages. The political standing of the unions in West Africa underwent an important change about 1960. While they had been in the van of nationalist struggles for independence in the late colonial period, once that objective was achieved they became interests in potential opposition to the national governments, and it was possible to represent them as organizations of relatively privileged groups of workers. Steps were then taken, especially in the francophone states and in Ghana, to remove the international affiliations of the unions, to place more severe legal inhibitions on their freedom of action, and to integrate them into the public administration or the organization of the ruling party. While such programmes to emasculate the unions were not invariably successful, taken with the unions' internal weaknesses they help to explain the movement of real wages traced above.

Nigeria, which has not so far been mentioned in the discussion of Table 13, is in some degree an exception to the generalizations that have been ventured. The federal constitution of the country, and its multiplicity of political parties before 1966 and since 1979, have militated against incorporation of the unions into the state apparatus, although some attempt in this direction was made by the military government in 1976–8. More or less organized movements of urban workers have appeared sufficiently threatening to governments for substantial wage increases to be conceded in 1964, 1970–1, 1975, 1980 and 1981.

Even so, the gains won by Nigerian wage-earners have been transitory in real terms. The real value of wages was eroded by rising prices in the later 1960s and, by the end of that decade, appears to have been less in all main centres of employment, or at best no more, than it had been in 1960.[66] According to the index used in Table 13, similar processes of erosion occurred in 1971–4, following the Adebo awards, and in 1975–9 after the Udoji Commission; the April 1979 figure in this index is, in fact, the same as that for 1964.[67] As to the gains won in 1980–1, it seems that about one-half had been lost through rising prices by the end of 1981.

Supposing it to be generally true both that minimum real wages in West Africa have been held down, or have improved only temporarily, since 1960, and that pay differentials have contracted in the same period, the conclusion would follow that real wages and salaries have tended to deteriorate throughout formalized employment. Such a fall could have been offset in some instances by improvements in fringe benefits. Places in formalized employment may also retain certain advantages even if real remuneration falls; they provide a relatively secure element in income, and the skills and personal contacts obtained through them may be important in developing other economic activities, to be carried on concurrently with or instead of wage-employment.

Summary

This chapter has examined some aspects of the human resources of West Africa. The size of the region's population is not known even roughly because of the politicization of census-taking in its most populous country. A United Nations estimate of 137 million in 1979, including 75 million Nigerians, is unlikely to be too low. The annual rate of population growth in the 1970s probably lay in the range of 2.5 to 3.5 per cent, both for the region as a whole and for most West African countries. Such rapid growth results from the coincidence of high birth rates (thought to average about 49 per thousand) and death rates (about 21 per thousand) which, though high by international standards, are now markedly below birth rates. Mortality is highly concentrated in the early years of life, so that life expectancy is both short by international standards – perhaps forty to fifty years – and usually higher at the age of five than at birth. The population is predominantly rural; urban population at the end of the 1970s was probably about one-fifth of the whole, though it has been increasing at least twice as fast as total population.

Because they are rapidly growing, the populations of West Africa are also youthful, with 40 to 50 per cent below the age of fifteen. Dependency ratios therefore appear high, but economic activity is by no means limited in practice to the population deemed to be of working age. Estimates of the relative size of the economically active population vary so considerably among West African countries as to make it improbable that real differences are being measured. Only small proportions are in what might be termed regular jobs, and consequently the distinction between those who are active and those who are inactive economically is bound to be arbitrary. Similar uncertainties arise in attempts to determine the distribution of the economically active population among economic sectors, though it is no doubt true that the majority of the labour force (55 to 80 per cent) finds its livelihood in agriculture and other primary activities in all West African countries, even the more commercialized.

Rapid population growth is usually postulated to impede improvement in labour productivity because it diverts resources from productive investment toward nurturing the young. This hypothesis takes for granted an effective demand for productive capital, but such a demand may be harder to create than the supply of investible resources, and population growth has historically played a part in helping to create it. The evidence of international and intertemporal comparison of estimated growth rates in population and in production is inconclusive in resolving the question whether population retards or helps propel economic growth, and it seems reasonable to conclude that the relationship between these variables is often weak. Those who stress that, on the contrary, population growth is an important (and negative) influence seem usually to have in mind its aggravation of problems in public administration rather than its effects on productivity.

Even if population growth were to detract from the increase of output per head, it might reasonably be regarded as a benefit received rather than a cost suffered, the more so since it results mainly from reduced mortality in infancy and early childhood and is associated with a lengthening expectation of life. Another non-economic dimension along which wellbeing may be improved is leisure, the enhancement of which may follow from reduction in intra-household services and from the extension of schooling. Of the latter, at least, measurements are available. Educational enrolments in West Africa as a whole rose absolutely from about 4.8 million in 1960 to about 16.4 million in 1977. In relative terms, net enrolments are likely to have risen from about 30 to 60 per cent of the relevant age-group in primary schools, and from about 3 to 12 or 15 per cent in secondary schools, between 1960 and 1980. Universal primary education could well be attained in the 1980s in Nigeria, the Ivory Coast and Togo. In the long run, a society progressing materially might be expected to increase its number, to enjoy greater output per head, and to use more of its time as leisure, at least so far as leisure is increased by the extension of schooling and deferment of entry to the labour force. All three forms of amelioration are occurring in West Africa, but the freedom to choose among them, and hence to maintain whatever might be regarded as an appropriate balance among population, output and leisure, is probably quite limited in practice.

Political concern in West Africa has been aroused less by the growth of population than by its shifting toward the larger towns, a movement often postulated to be out of phase with changing economic conditions and thus to lead to urban underemployment and possibly coincident shortage of rural labour. There are conceptual difficulties in positing a general excess of expected urban over rural earnings as the inducement of this migration, and the movement might be more realistically attributed to the sap of youth than to rational calculation. Higher urban wages could admittedly strengthen the inducement to migrate, not only directly but also indirectly through enlarging the capacity of the wage-earning community to carry dependants, but the same result could follow from increase in urban income-earning opportunities with wages constant or even (in real terms) falling – a combination of circumstances which, since 1960, has not been unusual in practice. The problem of the urban in-migration is not a problem to the migrant, who moves of his own volition or with the encouragement of his family, and the opportunity costs of the movement are for the most part borne privately. The aggravation of political and administrative difficulties by the migration does not necessarily denote reduction in the welfare of the people or the chances of economic progress.

Probably less than one-half of urban workers (and only some 5 per cent of the total labour force of West Africa) are in formalized employment in the so-called 'modern sector', but their importance is out of proportion with their number. Both the structure and the general level of their remuneration are politically sensitive. The highly elongated pay structures are not merely a survival of colonial practice but also reflect (especially in private employments) the relative scarcity of different grades of labour and degrees of educational attainment. In Nigeria, there is clear evidence that institutional forces (trade union pressure and the periodic general adjustment of pay following official wage commissions) have run counter to market forces over a long period by compressing pay differentials, especially in the public services; and it would not be surprising if this experience had been paralleled elsewhere in West Africa. On the other hand, institutional forces appear not to have produced, since about 1960, any sustained improvement in the pay of persons in formalized employment, and marked reductions in real wages appear to have occurred in some countries. The excess supply of labour generally believed to exist at the lower end of formalized employment has probably made itself felt, along with official hostility to autonomous trade unionism.

4

External Trade

The growth and changing composition of exports

It was said in Chapter 1 that overseas exports have been the prime mover in West African economic growth because distant markets have allowed more remunerative use of important West African resources than could be found locally. In particular, cultivable land, unskilled labour and mineral deposits have been either activated or made more productive in this way. Areas in the region able to exploit export outlets have generally prospered relatively to those in which such opportunities were lacking. The opportunities have not remained constant; they have changed with changes in tastes, technology and relative costs, and the ability, or failure, to respond to these shifts has affected the relative fortunes of areas in the region. The balance of private advantage between export and home markets has also moved, partly because of the growth of home demand impelled by export earnings, but also partly because of differing fiscal charges on exporting and on other activities.

The exporting bias in economic growth has produced difficulties which have been made much of in the literature expounding the political economy of West Africa (as of other developing areas) and which will be noticed in Chapter 6. For the moment, it is sufficient to observe that attempts to grow economically without an export bias would have also produced difficulties, and difficulties of greater magnitude, in regions so economically small and undiversified as West Africa.

The World Bank has emphasized the pernicious consequences of a slowing down in African export growth in recent years. In West Africa, the average annual growth rate in the volume of exports appears lower in 1970–9 than in 1960–70 in ten of the twelve countries for which estimates have been made (the exceptions are Mali and Niger), and that growth rate is estimated actually to have been negative in seven of the countries during the 1970s. The World

Bank regarded these changes in export quantum as deteriorating export performance, which it attributed to official discouragement of both exporting and agriculture, population pressure on agricultural resources, and the inflexibility of African economies which prevented their diversification into products with rapidly expanding markets.[1]

Changes in export volume are, of course, an adequate measurement of export performance only where the composition of the exports and the relative values of the export products remain fairly constant.

So far as export composition is concerned, there are West African countries in which little change has occurred since 1960. In Ghana, for example, the only significant changes have been the appearance of aluminium exports since 1967 and the relative decline of minerals other than gold; cocoa has regularly accounted for two-thirds or more of total export value, and timber and gold make up most of the remainder. Iron ore and crude rubber continue to preponderate in the exports of Liberia, and iron ore alone in those of Mauritania. In Sierra Leone, the main change has been the disappearance of iron ore exports since 1975; diamonds continue to account for one-half or more of total export value.[2]

Some diversification of exports has occurred in the Ivory Coast, and may be expected to continue in the 1980s, but the fall between 1963 and 1979 in the combined share of the three principal exports (coffee, cocoa beans and products, and timber) was only from about 85 to 75 per cent of total exports, as is shown in Table 14.

In Nigeria, on the other hand, the composition of exports has been transformed. Table 15 shows that agricultural products, consisting chiefly in cocoa, groundnuts and palm produce, made up about four-fifths of the total value of exports in 1960. Crude petroleum then became of fast-growing relative importance, and since 1974 its share of annual exports has averaged over 90 per cent. In 1980, the old agricultural export staples contributed only some 2.5 per cent of the export total. Even sharper was the change in Niger; in 1970, the export of uranium had barely begun and groundnuts made up nearly three-fifths of the export total, while in 1975 uranium ore was as relatively important as groundnuts had been five years earlier, and groundnuts now comprised less than 1 per cent of the export total.

The transformation in Nigeria is the more significant because of the large and growing importance of the country in the exports of

Table 14 Composition of exports: Ivory Coast, 1963, 1971 and 1979

	1963	*1971*	*1979*
	Value of exports (bill. CFA fr.)		
	74.5	126.6	534.8
	Percentages of total export value		
Coffee: green	43.1	33.3	31.1
soluble		1.4	1.3
Cocoa: beans	19.9	17.1	21.7
products		3.5	5.5
Timber: logs	21.8	20.5	12.6
processed		4.0	3.3
Three principal commodities	84.8	79.8	75.5
Bananas	6.2	2.3	0.7
Pineapples, fresh and canned	1.8	3.6	1.3
Palm produce	0.5	2.0	1.8
Kola nuts	0.7	0.7	0.2
Fish, canned	–	0.5	1.5
Rubber	–	0.9	0.8
Cotton	–	1.4	1.9
Sugar	–	–	0.2
Diamonds	0.7	0.5	–
Manganese ore	0.7	–	–
Petroleum products	–	0.5	4.5
Cotton textiles	–	–	2.1
Other products	4.6	7.8	9.5
	100.0	100.0	100.0

Sources: Surveys of African Economies, vol. 3 (Washington DC: International Monetary Fund, 1970), Table 32, p. 297; Banque Centrale des Etats de l'Afrique de l'Ouest, *Notes d'information et statistiques*, août–septembre 1976 and février 1982.

West Africa as a whole. In 1960, Nigeria provided 36 per cent of the total dollar value of West African exports. Ghana was next in importance with 22 per cent. In 1979, the Nigerian share was over 74 per cent. Nigeria and the Ivory Coast together contributed 86 per cent. The Ghanaian share was now less than 5 per cent. The other twelve countries of mainland West Africa together accounted for less than 10 per cent of the total.[3]

It is true that the exports of West Africa are still preponderantly unprocessed or simply processed agricultural and forest products and minerals – cocoa, coffee, groundnuts, palm kernels, vegetable

Table 15 Composition of exports: Nigeria, 1960, 1970 and 1980

	1960	*1970*	*1980*
	Value of exports (million naira)		
	339	877	14,077
	Percentages of total export value		
Cocoa beans	21.7	15.2	2.2
Cocoa products	–	1.8	0.2
Groundnuts	13.5	5.0	–
Groundnut oil	3.1	2.6	–
Groundnut cake	–	1.3	–
Palm oil	8.2	0.1	. .
Palm kernels	15.4	2.5	0.1
Cotton	3.7	1.5	. .
Hides and skins	2.5	0.6	. .
Rubber	8.4	2.0	0.1
Timber, logs and sawn	4.1	0.7	. .
Tin metal	3.6*	3.9	0.1
Crude petroleum	2.6	58.2	96.3
Other domestic products	13.2	4.8	1.0
	100.0	100.0	100.0

Notes: . . indicates negligible.
* tin ore.
Data for 1980 are provisional.
Re-exports excluded.

Sources: F. O. Fajana, 'International Trade and Balance of Payments', in F. A. Olaloku *et al.*, *Structure of the Nigerian Economy* (London: Macmillan, 1979), Table 11.3, p. 228; Central Bank of Nigeria, *Annual Reports*.

oils, rubber, cotton, timber, iron ore, diamonds, uranium ore, phosphates, crude petroleum. But whereas minerals made up less than one-fifth of the total value of West African exports in 1960, their contribution by 1979 had reached about three-quarters, mainly because of increase in the volume and prices of Nigerian petroleum exports.

The quantum indices of West African exports provide a poor indicator of export performance, not only because the composition of the region's exports changed profoundly in the twenty years following 1960 but also because of changes in the relative values of the exports.

Changes in the values of exports relatively to imports are

measured by the commodity, or net barter, terms of trade (the ratio of an export price index to an import price index). Multiplying these commodity terms by an export volume index gives the income terms of trade, a measurement of changes in the overall purchasing power of exports. A fall in export volume therefore counteracts the effect on the income terms of improving commodity terms (resulting from dearer exports, or cheaper imports). Conversely, a rise in export volume would reinforce the effect on the income terms of improving commodity terms.

Table 16 is derived from tentative estimates of the terms of trade made by UNCTAD for thirteen of the West African countries. The base year is 1975. In order to reduce the significance for comparisons of the choice of particular years, average index numbers are shown in the table for each of the three-year periods 1960–2, 1969–71 and 1977–9. The percentage rates of annual growth in the indices are then shown for the two periods, 1960–2 to 1969–71 and 1969–71 to 1977–9.

It will be observed that, in the first period, the commodity terms improved, albeit sometimes only modestly, in all the countries except the two iron ore exporters, Liberia and Mauritania. There was more variety of experience in the second period. The deterioration continued in Liberia and Mauritania and was experienced in less degree in six other countries. On the other hand, there are several instances of rapid improvement in the commodity terms: above all, Nigeria (because of the petroleum price increases in 1973–4 and 1979), but also Ghana, the Ivory Coast and Togo.

The income terms improved for all the countries, including the iron ore exporters, during the first period. If an annual improvement in these terms of, say, 5 per cent were deemed concomitant with a briskly growing economy, only The Gambia, Ghana, Senegal and Sierra Leone could be said to have lagged. In the second period, more diversity again appears. The purchasing power of exports increased very rapidly in Nigeria, and also markedly in the Ivory Coast, Mali, Niger and Togo, but it was almost constant in Ghana and it actually fell in Benin, Sierra Leone and the countries exporting iron ore.

Trends in these terms of trade estimates for Nigeria, Ghana and the Ivory Coast are plotted on the basis of three-year moving averages in Figure 1, showing the commodity terms of trade, and Figure 2, showing the income terms. For Nigeria, the outstanding feature in the movement of the commodity terms is the vast

Table 16 Evolution of commodity and income terms of trade of West African countries, 1960–79

	Av. 1960–2	Av. 1969–71	Av. 1977–9	Annual growth rate (%) 1960–2 to 1969–71	1969–71 to 1977–9
A. COMMODITY TERMS OF TRADE (1975=100)					
Benin	110	127	104	1.6	−2.5
The Gambia	102	109	107	0.7	−0.2
Ghana	99	107	164	0.9	5.5
Ivory Coast	106	119	150	1.3	2.9
Liberia	219	134	95	−5.3	−4.2
Mali	106	116	108	1.0	−0.9
Mauritania	151	129	84	−1.7	−5.2
Niger	97	108	98	1.2	−1.2
Nigeria	30	32	110	0.7	16.7
Senegal	70	79	91	1.4	1.8
Sierra Leone	118	135	115	1.5	−2.0
Togo	51	56	97	1.0	7.1
Upper Volta	83	116	103	3.8	−1.5
B. INCOME TERMS OF TRADE (1975=100)					
Benin	104	213	79	8.3	−11.7
The Gambia	43	59	79	3.6	3.7
Ghana	89	100	103	1.3	0.4
Ivory Coast	37	82	155	10.4	8.3
Liberia	44	115	99	12.8	−1.9
Mali	52	96	175	7.0	7.8
Mauritania	3	96	65	47.0	−4.8
Niger	40	66	136	6.5	9.5
Nigeria	13	31	124	11.5	18.9
Senegal	60	62	79	0.4	3.1
Sierra Leone	131	159	89	2.2	−7.0
Togo	31	79	126	11.0	6.0
Upper Volta	23	79	98	16.7	2.7

Source: Calculated from tentative terms of trade estimates in UNCTAD, *Handbook of International Trade and Development Statistics, Supplement 1980* (New York: United Nations, 1980), Table 7.2.

improvement between 1972 and 1975, resulting from the quadrupling of the petroleum export price during that period. The trend in the commodity terms was similar for Ghana and the Ivory Coast, though the amplitude of movements was less for the Ivory Coast and it was spared the deterioration experienced by Ghana in the early 1960s. Both countries met with improving terms in the late 1960s, deterioration in the early 1970s, and a phase of rapid improvement after 1975.

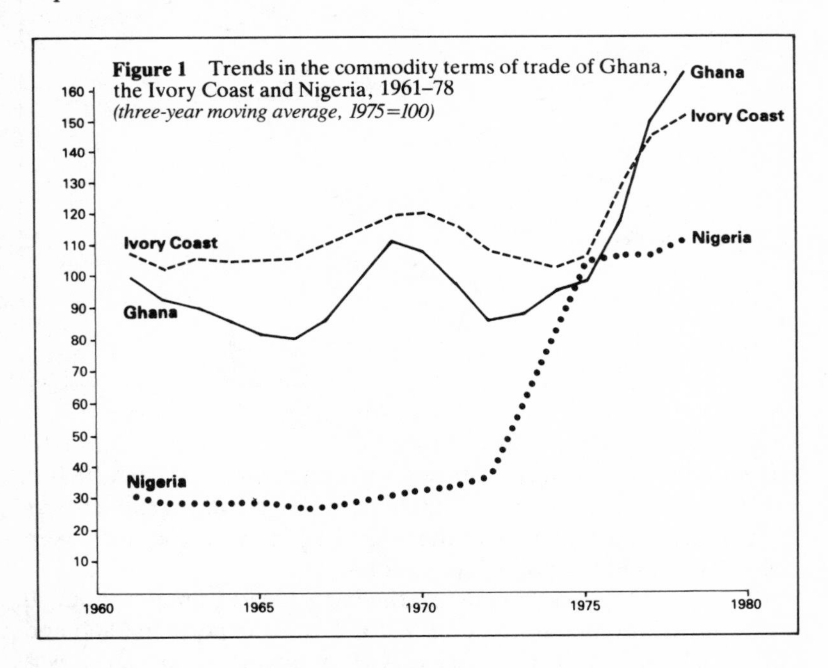

Figure 1 Trends in the commodity terms of trade of Ghana, the Ivory Coast and Nigeria, 1961–78
(three-year moving average, 1975=100)

Turning to Figure 2, the upward trend in the Nigerian income terms is shown to have begun at the beginning of the 1970s, when it reflected increased volumes of exports (particularly petroleum) rather than rises in their relative prices, and to have accelerated greatly after 1972. By the mid-1970s, the purchasing power of Nigerian exports over imports was apparently some sixfold what it had been before the civil war of 1967–70. As between Ghana and the Ivory Coast, the movement in the income terms was very different, in contrast to the similarity of their experience of the commodity terms. For Ghana, there was some modest improvement (about 25 per cent) in the annual average purchasing power of

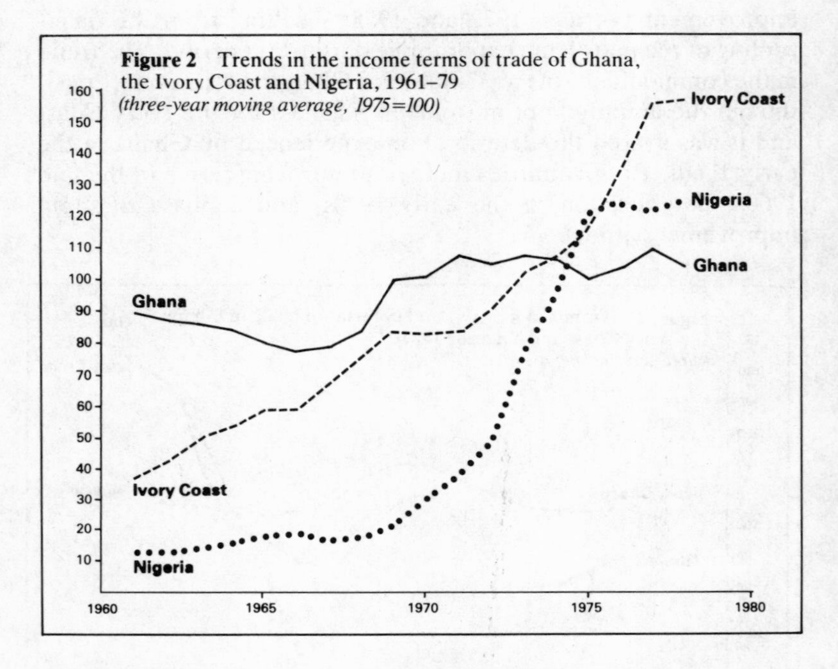

Figure 2 Trends in the income terms of trade of Ghana, the Ivory Coast and Nigeria, 1961–79 *(three-year moving average, 1975=100)*

exports between the 1960s and the 1970s. For the Ivory Coast, the upward trend in the income terms is almost continuous from 1960 to 1978, and, by the latter date, the purchasing power of exports was fourfold what it had been at the beginning of the 1960s, such were the increases in the volume of exports.

Together, Nigeria, Ghana and the Ivory Coast supplied about 70 per cent of the dollar value of West African exports in 1960 and about 90 per cent in 1979. West African export performance was therefore strongly influenced by what happened in those three countries, and more especially in Nigeria and the Ivory Coast, since Ghana was of fast-diminishing relative importance as an exporter. For West Africa as a whole, then, the trend in the commodity terms of trade was mildly favourable between 1960 and 1972 and highly favourable from 1973 to 1979, while the purchasing power of exports rose rapidly throughout the period and especially from the end of the 1960s.

Of course, not all the smaller West African countries shared this overall experience. And the years immediately following 1979 were much less favourable for even the larger countries. The Ivorian commodity terms of trade, which had peaked about 1978, continued to decline and in 1981 may have been below the level of 1975. The income terms of trade must have deteriorated too. The same was probably the case with Ghana. In Nigeria, the commodity terms must have continued to improve, since the average export price of petroleum was raised some 80 per cent between 1979 and 1981, but the volume of petroleum exports was falling and the purchasing power of exports may therefore have been little or no greater in 1981 than in 1979.

Even so, it would clearly be inconsistent with the facts to suggest that the West African region had suffered the consequences of a retardation in export growth since, say, 1970. Changes in the composition of exports and in the relative prices of export products have allowed foreign trade to continue as a strongly propellent force in the economic growth of the region – albeit some particular countries, such as Ghana, Liberia and Sierra Leone, have lost this momentum through causes internal or external to themselves.

Changes in import composition

Growth in the value of West African imports, and shifts in their distribution among countries, have naturally been determined mainly, though not entirely, by the development of the region's export trade. Measured in current US dollars, West African imports rose from $1,558 million in 1960 to $2,734 million in 1970 and $16,470 million in 1979. Table 17 shows that Nigeria and Ghana were the principal importers in 1960, taking together over three-fifths of the West African total; that the Ivory Coast was of almost equal importance to Ghana by 1970; and that Nigeria and the Ivory Coast took together nearly three-quarters of the total in 1979.

The composition of the imports has changed over time – because of the changes in the structures of production and expenditure in West African countries that were outlined in Chapter 2, and to a lesser extent because of changes in import costs, particularly the rise in petroleum prices from 1973.

Table 18 shows the changing relative importance in each of thirteen West African countries of imports of food, fuel and machinery and transport equipment. Food is defined as the sum of Sections 0 (food and live animals), 1 (beverages and tobacco) and 4 (animal and vegetable oils and fats) and Division 22 (oilseeds, oil nuts and oil kernels) of the Standard International Trade Classifica-

Table 17 Country shares in West African imports, 1960, 1970, 1979

	1960	*1970*	*1979*
	(value of West African imports in millions of current US dollars)		
	1,558	2,734	16,470
	(percentages of West African imports)		
Ghana	23.3	15.0	6.0
Ivory Coast	7.7	14.2	15.1
Liberia	4.4	5.5	1.4
Nigeria	38.8	37.7	58.8
Senegal	7.8	7.1	5.8
Sierra Leone	4.7	4.2	1.6
Other countries of mainland West Africa	13.3	16.3	12.4
	100.0	100.0	100.0

Source: UNCTAD, *Handbook of International Trade and Development Statistics, Supplement 1980* (New York: United Nations, 1980), Table 1.2.

tion; fuel is SITC Section 3 (mineral fuels and lubricants); and machinery and transport equipment are Section 7. In each country, the remaining imports include crude materials other than fuel (Section 2) but are mainly made up of other manufactures (Sections 5, 6 and 8).

The ratios of food imports appear high for countries that find so much of their GDPs and use so much of their labour in agriculture and related activities, but they do not appear to have trended upward (except in Mauritania) and there are several instances (including Ghana, the Ivory Coast and Senegal) of decline. (In absolute terms, food imports have grown nearly everywhere, but the same could be said of any other broad category of import goods.)

The effect of the petroleum price increases in 1973–4 is seen in a doubling or trebling in the relative importance of fuel imports in many countries between 1970 and 1978. The further phase of petroleum price increases in 1979–80 will have raised these ratios still further. On the other hand, fuel imports have declined in relative importance in Nigeria as the country's petroleum industry has been developed. (They have not disappeared entirely because

Table 18 Relative importance of food, fuel and machinery and transport equipment in imports of West African countries, selected years, 1960–78

(percentages of total imports)		*Food*	*Fuel*	*Machinery and transport equipment*
Benin	1960	17	10	18
	1970	18	4	21
	1978	15	15	22
The Gambia	1967	22	4	19
	1970	32	4	15
	1975	24	9	14
Ghana	1960	19	5	26
	1970	21	6	26
	1977	9	16	26
Ivory Coast	1960	18	6	27
	1970	16	5	33
	1978	13	10	39
Liberia	1960	16	4	34
	1970	17	6	34
	1978	17	18	32
Mali	1967	18	8	22
	1970	29	9	21
	1978	19	14	30
Mauritania	1967	17	4	53
	1970	23	8	38
	1975	31	8	35
Niger	1967	16	6	20
	1970	14	4	26
	1978	10	12	33
Nigeria	1960	14	5	24
	1970	8	3	37
	1978	14	2	44
Senegal	1960	30	5	19
	1970	29	5	25
	1978	23	12	18
Sierra Leone	1960	23	12	15
	1970	26	5	26
	1978	21	12	24
Togo	1960	16	6	32
	1970	23	4	22
	1978	8	14	37
Upper Volta	1967	27	7	17
	1970	20	8	27
	1978	19	9	43

Sources: Accelerated Development in Sub-Saharan Africa (Washington DC: World Bank, 1981), Statistical Annex, Table 9; UNCTAD, *Handbooks of International Trade and Development Statistics* (New York: United Nations), 1976, Table 4.1; Supplement 1980, Table 4.2.

refining capacity lagged behind the growth of home demand for petroleum products in the 1970s, and because the light crudes extracted in Nigeria are not suitable for the production of all petroleum products.)

The import ratios for machinery and transport equipment are high, reflecting the lack of engineering and capital-goods producing capacity in West Africa and, in some instances, the maintenance of very high investment rates. An upward trend in the relative importance of this category of imports is clearly suggested by the figures for the Ivory Coast and Nigeria.

For West Africa as a whole, changes in import composition are, of course, strongly influenced by those two countries, especially the latter. It can be roughly estimated that food fell as a proportion of all West African imports from about 17 per cent in 1960 to 16 per cent in 1970 and 14 per cent in 1978. The fuel ratio remained fairly stable, since increases in many countries were offset by the decline in Nigeria; the percentages are roughly 6 in 1960, 5 in 1970 and 6 in 1978. This ratio can be expected to trend down in the early 1980s, with the emergence of an Ivorian petroleum industry. The regional ratio for machinery and transport equipment rose strongly – from about 24 per cent in 1960 to 32 per cent in 1970 and 40 per cent in 1978. The residual category, consisting mainly in other manufactures, consequently contracted from about 53 per cent in 1960 to 47 per cent in 1970 and 40 per cent in 1978.

Accompanying the changes in the commodity composition of imports have been shifts in their uses for consumption, production and capital formation. Table 19 presents comparisons for two countries which are important both in themselves and as alternative patterns of change. In Nigeria, during the period of rapid economic growth between the last years of the first Republic (1963–5) and the end of the 1970s, in which purchasing power over imports rose about eightfold, there were offsetting and almost equal movements in the relative importance of consumers' goods and capital goods; the former declined from nearly one-half of total imports to just over one-quarter, while the latter rose from just over one-quarter to nearly half. A large fraction of the increments in GNP was passing through official hands and much of it was being used in investments requiring imported equipment. On the other hand, raw materials and intermediate products showed little change in relative importance between the periods that are compared, although they increased greatly in absolute value.

Table 19 End-use analyses of imports: Nigeria and Ghana, selected periods (percentages of total imports)

A. NIGERIA

		1963–5		1978–80
Non-durable consumers' goods		39.2		20.6
Durable consumers' goods		8.1		7.9
		47.3		28.5
Raw materials		23.7		22.9
Capital goods		27.6		48.5
Miscellaneous		1.4		0.1
		100.0		100.0

B. GHANA

		1956–8		1969–71
Non-durable consumers' goods				
Food, drink and tobacco	16.1		12.1	
Textiles and cloth	19.6		5.2	
Other	9.8		6.8	
		45.5		24.1
Durable consumers' goods				
Private vehicles & accessories	3.4		3.0	
Other	6.3		2.3	
		9.7		5.3
		55.2		29.4
Raw and semi-finished materials				
For food, drink and tobacco	5.9		6.2	
For agriculture	2.1		1.5	
For mining and manufacturing	5.8		20.9	
For construction	11.1		10.9	
		24.9		39.5
Capital equipment		14.0		25.0
Fuels and lubricants		6.0		6.0
		100.0		100.0

Sources: Central Bank of Nigeria, *Annual Reports*, 1965, Table 24; 1980, Table 57; *Economic Surveys* (Accra: Central Bureau of Statistics), 1965, Table 12; 1969–71, Table 3.3.

Ghana is a contrasting case. At best, economic growth has been slow in recent times. The purchasing power of exports increased only modestly (by about 25 per cent) between the periods shown in the table. Yet the end-use analysis of imports again shows marked changes between 1956–8, the years immediately preceding important changes in official development policies, and 1969–71, by when the resulting economic transformation had been largely completed. Between these periods, the relative importance of consumers' goods fell from over one-half of total imports to only three-tenths. The contraction was especially marked in textiles and cloth, a long-established staple of the West African import trade. The offsetting increases were partly in capital goods, but more largely in materials and semi-products, especially those used in manufacturing industry. The causal factor was a deliberate diversification of the economy undertaken with government finance and borrowings from abroad and sustained through the administrative diversion of import spending from consumers' goods to industrial inputs and equipment.

Economic growth and economic diversification therefore underlie changes in the composition of imports. While the latter is a natural outcome of the former, it can also occur independently of economic growth, and even at the expense of economic growth. In any event, changes will be produced in the structures of production and expenditure, and they will find reflection in the composition of imports.

Changes in the direction of trade

Shifts in the direction of West Africa's external trade have occurred partly as a result of the changes in the composition of exports and imports discussed above, and partly because of changes in commercial policy.

Although the external trade of British West Africa had been much liberalized in the closing years of colonial administration, the United Kingdom still supplied 42 per cent of Ghana's imports in 1957, the year of independence, and took 37 per cent of the country's exports. The corresponding percentages for Nigeria in 1960 were 42 and 48 respectively. The importance of France in the trade of French West African dependencies was even greater. In 1960, France provided 70 per cent of the Ivory Coast's imports and took 53 per cent of her exports. In Senegal, 69 per cent of the imports came from France and 82 per cent of the exports went to that destination. The trade statistics even reveal the affiliation of the independent state of Liberia with the USA; in 1960, just over one-half of both the import and export trade was with America.

All these ratios had been greatly reduced by the end of the 1970s. In 1977, the UK provided only 14 per cent of Ghana's imports, and took 19 per cent of her exports. In 1980, 22 per cent of Nigerian imports came from the UK, and only 2 per cent of Nigerian exports went to that destination. France supplied 37 per cent of Ivorian imports in 1979 and took 24 per cent of Ivorian exports. The proportions for Senegal were higher – 40 and 46 per cent respectively – but still well below the levels of 1960. The USA provided 24 per cent of Liberia's imports in 1979 and took 20 per cent of her exports. In all the other countries of West Africa, the predominance of the principal trading partner had been greatly reduced since 1960.[4]

The external trade of West Africa is still very largely directed toward the industrialized market economies of Western Europe (especially the European Community), the United States and Japan. Even so, the diversification of trading connections is considerable by the standards of earlier times, especially in the import trade. By the late 1970s, countries other than the five principal importing partners provided over 40 per cent of the total imports of Benin, The Gambia, Ghana, the Ivory Coast, Liberia and Senegal, and over 30 per cent of those of Mauritania, Niger, Nigeria, Sierra Leone, Togo and Upper Volta.

Political autonomy has undoubtedly played a part in producing these changes. The markets of the independent West African states have become more widely accessible than were those of the colonial territories. The responsible governments have been readier to encourage low-cost sources of supply and to discover new sources of import credit. In a few countries, inter-governmental agreements have made for abrupt changes in trading patterns; thus, the share of Russia and Eastern Europe in the total trade of Ghana was raised from 5 to 24 per cent between 1960 and 1965. The association of African countries with the EEC under the Treaty of Rome and the later Lomé Conventions has facilitated diversification of West African trade among West European countries, and has particularly affected the francophone West African states.

Of by far the greatest importance as a determinant of the direction of export trade has been the domination of Nigerian exports by petroleum since the end of the 1960s. In 1980, nearly 60 per cent of Nigeria's exports of crude went to refineries in the USA and the Caribbean, and another 30 per cent was shipped to the Netherlands, France and West Germany. The UK became relatively insignificant as a Nigerian export market because the British were themselves extracting light crudes from the North Sea oilfields, and because Nigerian exports other than oil had shrunk to only a few per cent of total exports.

Because of Nigerian oil, the export ties of West Africa have become as much transatlantic as with the European Community (in each case, about 43 per cent of 1979 exports). The import ties are more largely with the European Community (about 56 per cent of 1979 imports). As was mentioned in Chapter 1, international trade within West Africa is relatively small – less than 3 per cent of recorded exports in 1979. Only for Guinea-Bissau, Mali and Senegal were exports to other African countries substantial (over 15 per cent of total exports) in 1979; only for Mali, Mauritania and Upper Volta were imports from African countries substantial.

External payments and debt

As with so many other West African economic data, the records of the payments generated by external trade and other international transactions are highly imperfect. It was said of the countries of the Sahel in 1975 that: 'everything related to external trade – merchandise exports and imports, remittances, the balance of payments as a whole – is necessarily part fantasy, since so much of total trade and total intra-regional economic relationships takes place unrecorded.'[5] It can be added that transactions often escape record because they are illegal, and that the volume of illegal dealings has probably increased in recent years – as with Ghanaian trade with Togo and the Ivory Coast, and the trade of Benin with Nigeria. The structure of a country's published balance of payments may therefore be illusory – a large recorded trade deficit may, for instance, be offset by unrecorded re-exports – and year-to-year movements in the published figures may be difficult to interpret. Even so, the general character of the external payments and receipts of West African countries, and the changes that have occurred over time, are tolerably clear.

Thus, as was mentioned in Chapter 2, the territories of British West Africa maintained substantial export surpluses in the decade following the Second World War, when their export earnings were buoyant, much of these earnings passed into official hands through tax systems strongly connected with external trade and through export marketing monopolies, and there was a lag in the public capacity to spend money usefully. Nigeria, the Gold Coast and Sierra Leone together had recorded trade surpluses totalling $1,143 million (current dollars) in the period 1945–54. In contrast, in the years 1955–60, these three territories had trade deficits totalling $462 million; export proceeds had become less buoyant after 1954, while the public capacity to spend had increased, or had become less inhibited. The trade deficits of this later period, along with the regular deficits in invisibles (services and factor payments) and some capital outflow in the late 1950s, were met substantially by drawing on the external assets accumulated in the earlier period.

The French territories together had trade deficits throughout the late colonial period; over the years 1945–60, they totalled $1,005 million. These deficits were financed by – and, indeed, were occasioned by – French government grants and loans, which also covered deficits in invisibles and considerable outflows of private capital. In 1958–60, the years immediately preceding independence, French official transfers to AOF and Togo were running at about $200 million per year.[6]

In formerly British West Africa, the running down of surplus external assets (including those supposedly accumulated in order to stabilize export-crop farmers' income) continued in the early years of independence. This process was virtually completed in Ghana by 1962, in Nigeria by 1964, and in Sierra Leone by 1966.

An important part in allowing current payments deficits to continue was then played by the provision of medium-term credits by foreign suppliers and contractors. The life of these credits ranged between five and twelve years, but on average was not much above the lower limit. It was a form of external finance that quickly became scandalous. Although ostensible interest rates were moderate, heavy borrowing costs were added by other financing charges and the frequent overpricing of equipment supplied. Neither the borrowers nor the lenders (and the official institutions in the creditor countries that frequently insured the lenders' risks) were adequately concerned that the credits should be used only in investments likely to produce rapid rates of return, commensurate with the short life of the credits. There was similar neglect of a borrowing country's likely overall capacity for debt servicing. There was even ignorance of the amount of debt that was being incurred in each country. Obtaining credits usually owed as much to the enterprise of salesmen as to the financial needs of the borrowing ministries and public corporations, and was attended by corruption as well as fraud.[7]

Where payments deficits exceeded such financing resources, and reserves were almost exhausted, there was resort to short-term borrowing from abroad, including bank loans and the deferment of payment commitments. In Ghana, the government made 180-day import credits mandatory in 1965. As with the medium-term suppliers' credits, such stratagems were of immediate value in allowing import surpluses to be maintained, but only at the cost of raising the prices of imports and exacerbating future payments difficulties.

About two-fifths of Ghana's imports in the period 1961–5 were financed by the depletion of reserves and by commercial credit.[8] At the time of the military coup in February 1966, some £40 million ($112 million) was due in short-term external debt commitments, and the outstanding medium-term debt totalled £151 million ($423 million). The magnitude of these debts was such that, if servicing and repayment commitments had been honoured in 1966, they would have absorbed about one-half of the current annual value of exports.[9] In the event, the new government declared a moratorium, and re-scheduling of the medium-term debts followed, by agreement with the creditors, at the end of 1966 and in 1968 and 1970. The Ghanaian government thus received successive phases or tranches of relief, pushing the bulk of the debt forward into the 1970s but at the cost of increasing its amount, since interest was charged on these deferments. Finally, in 1974 a re-financing agreement was made, whereby the governments of the creditors replaced the medium-term debts with long-term, low-interest loans to the government of Ghana, payments on which were not to begin until 1983. The greater part of the cost of the suppliers' credits extended to Ghana in the early 1960s was thus eventually borne by the taxpayers of the creditor countries.

The government of Sierra Leone underwent a similar, though less acute, phase of embarrassment by external debt in the mid-1960s, and had to seek the support of the IMF at the end of 1966.[10] Again, medium-term suppliers' credits, accepted for reasons which often did not bear close inspection, had a prominent place in building the debt. The same pattern was unfolding in Nigeria,[11] and its consequences were avoided only through the effects of the emerging petroleum industry on foreign exchange receipts and public revenues.

External indebtedness also grew rapidly about this time in Liberia, Guinea and Mali. Because the currency of Liberia is the United States dollar, the debt appeared there as a budgetary rather than a balance-of-payments problem.[12] A steep rise in budgetary receipts in the 1950s, arising from growth in rubber and iron ore production and the import trade, had encouraged the government to borrow heavily for public works at the beginning of the 1960s, much of the finance being provided by European contractors. As was to happen in Ghana, the debt snowballed for want of central control. By 1963 it totalled $123 million, including $61.5 million in contractor finance and $15 million in short-term bank credit. Debt charges over the period 1963–7 would have been payable at an annual average rate of over $20 million, and in 1963 the amount due was $33.5 million – over 90 per cent of the government revenue estimated for that year. The debts were rescheduled in 1963. Over the next decade, between 20 and 30 per cent of annual government spending was devoted to their servicing.

Guinea and Mali were the two countries of the former AOF to assert monetary autonomy soon after their political independence by issuing national currencies – the former in 1960 and the latter in 1962. The governments undertook ambitious programmes of investment and institutional reform on the basis of inadequate foreign exchange reserves and soon became heavily indebted, with Russian and East European suppliers figuring more prominently among the creditors in these cases. In Mali, the disbursement of foreign loans by mid-1966 totalled about $120 million, equivalent to nearly one-third of the GDP estimate, and 'the debt service obligations accumulated for the future were staggering.'[13] As in Ghana, Liberia and Sierra Leone, the assistance of the IMF had to be solicited and the terms of a stabilization programme agreed;[14] the government also negotiated with France an agreement to restore the convertibility of Malian currency.

The difficulties in making external payments, and the associated burdens of external debt, that appeared so soon after independence in several West African countries might be partly attributed to an unexpected slowing-down in export growth, at least in Ghana and Sierra Leone[15] (see Table 16B and Figure 2). It might also be argued that international aid in support of these fledgling states was inadequate – though it could be added that this would hardly be true of Guinea, that the Ghanaians were ambivalent in their attitude toward aid and made little attempt to attract it, and that aid fell short of expectations in Nigeria because of a lack of acceptable projects on which the commitments offered by donors could be disbursed.[16] More important as sources of payments difficulties were two other factors: the abandonment with which the governments concerned embarked on new investments, finding outlets for funds that were frequently untenable by any economic criteria and consequently having to bear all or most of the risks themselves; and the pressure of domestic demand on the balance of payments in those cases (notably Ghana, Guinea and Mali) where credit creation through an independent monetary system was being used to allow government domestic expenditures much in excess of revenues. Unreasonably ambitious government plans and unreasonably high government expenditures had more to do with these payments difficulties than did unreasonably low receipts of foreign exchange.

Other West African countries fared better in the 1960s. They were more careful to maintain the relationships of financial support by the former metropolitan power that had been built in the last phase of colonialism.[17] For some, like the Ivory Coast, and The Gambia after 1964, export earnings continued to be buoyant. All of them retained monetary dependence and, in consequence, monetary sobriety. The Gambia was the one surviving member of the old West African Currency Board until 1971, when it acquired a central bank.[18] Dahomey, the Ivory Coast, Mauritania, Niger, Senegal, Togo and Upper Volta became members of a monetary union, the *Union Monétaire Ouest Africaine* (UMOA), and continued to use a common currency (the CFA franc) issued by a common central bank, the *Banque Centrale des Etats de l'Afrique de l'Ouest* (BCEAO).[19] The foreign exchange reserves of the UMOA were kept in an operations account maintained by the BCEAO with the French Treasury, in return for which the French guaranteed the unlimited convertibility, at least into French francs, of the West African CFA franc. Monetary discipline in the UMOA rested ultimately on French participation in the government of the common central bank, and on the common interest of members of the union in the prevention of excessive credit creation in any one of their number, given that they had a currency in common.

In the 1970s, the balances of payments of the West African countries were powerfully affected by the oil price increases in 1973–4 and at the end of the decade, by droughts in the Sahelian areas, by enormous fluctuations in the prices both of export-crops and of imported foodstuffs, and by economic recession in the industrialized countries of the Western world. West African governments were among the beneficiaries as well as the victims of this commotion, and they added to it tumults of their own devising, either by persistent discouragement of exports (as, notably, in Ghana) or by overreacting to export booms (as in the Ivory Coast, Nigeria and Togo).

The improvement in the Nigerian balance of payments through oil exporting was retarded by the civil war of 1967–70. In that period, the government used exchange control to defer external payments that were due, and later it made short-term credits mandatory for many imports; not until 1974 was the backlog of payments altogether cleared. Meantime, the success of the inter-governmental Organization of Petroleum Exporting Countries in wresting control over crude oil pricing from the producing companies allowed the Nigerian posted price[20] to be raised from an average of $3.39 per barrel in 1972 to $14.69 in 1974, and the value of Nigerian exports increased between these years from less than $2.2 billion to $9.7 billion, as is shown in Table 20. Both the current and the overall balances showed surpluses approaching $5 billion in 1974, and the foreign exchange reserves at the end of that year would have bought two years' worth of imports at the current rate.

The Nigerian authorities reacted impetuously to this astonishing transformation in economic prospects. Between 1974 and 1976 the third national plan was enlarged four or fivefold in real terms.[21] The actual total spending of the federal government increased from ₦3 billion to ₦8 billion between the same years. The flow of imports was so increased in 1975 that delays in berthing at Nigerian ports ran into months and a queue of several hundred ships formed in the Lagos roadstead. By 1976, the balance of payments was in deficit. In 1978, when export proceeds were checked by a fall in oil sales, a negative trade balance appeared and the current deficit was over $3.7 billion. In that year, the government began raising syndicated medium-term loans from banks in the Eurocurrency market; the scale of this and other official borrowing in 1978–80 can be seen in Table 20. Also in 1978, a comprehensive import supervision scheme was prepared, making shipments subject to inspection to verify

Table 20 Nigeria: balance of external payments, 1972–80 (US $m)

	1972	*1973*	*1974*	*1975*	*1976*	*1977*	*1978*	*1979*	*1980*
Exports (fob)	2,184	3,602	9,711	8,312	10,123	12,376	10,444	16,769	23,422
LESS imports*	1,459	1,829	2,650	5,896	8,059	9,965	11,610	11,844	15,948
Trade balance	725	1,774	7,061	2,416	2,063	2,411	−1,167	4,924	7,474
Services and income (net)	−1,195	−1,640	−2,091	−2,222	−2,322	−3,231	−2,321	−2,861	−3,982
Unrequited transfers (net)	−21	−53	−99	−126	−156	−183	−269	−387	−577
Current balance	−491	81	4,871	68	−415	−1,003	−3,757	1,676	2,915
Private capital transactions (net)	354	296	94	450	294	259	351	363	470
Official capital transactions (net)	55	−76	−104	−221	−373	−26	1,400	987	948
Errors and omissions	6	−35	73	−42	−49	−47	−30	76	64
Overall balance	−76	266	4,935	255	−543	−818	−2,036	3,103	4,397

Notes: Naira totals have been converted to US dollars at annual average rates of exchange.
* Imports valued cif until 1976, fob from 1977.
Data for 1980 are provisional.

Source: Central Bank of Nigeria, *Annual Reports*.

quantities, qualities and prices, with a view to arresting the inflow.

There would have been much less alarm about the Nigerian balance of payments in 1978 if it had been possible to foresee the Iranian revolution at the end of that year and its effects on world oil supplies and prices. In the event, it was possible to increase the volume of Nigerian oil production by one-quarter in 1979, compared with the previous year, and the export price[22] of Nigerian oil was raised from an average of about $14 per barrel in 1978 to a peak of $40 at the beginning of 1981. The resulting surge in export proceeds, supported by import restrictions in 1979 and by further public borrowing from abroad, produced large overall surpluses in 1979–80, as can be seen from the table, and foreign exchange reserves exceeded $10 billion at the end of 1980.

The Nigerian authorities responded with alacrity to this good fortune, just as they had five years earlier. Total federal outlays were almost doubled (from ₦7.6 billion to ₦14.8 billion) between the fiscal years 1978/79 and 1979/80. Expenditure (public and private) on fixed capital formation was estimated as ₦11.8 billion in 1980, about double the average of the preceding four years. The total of public investment planned for 1981–5 was almost trebled (from ₦24.4 billion to ₦70.5 billion) between the preparation of guidelines for the fourth national plan in 1978 and publication of an outline of the plan at the beginning of 1981.

Once again, the euphoria was quickly dispelled. Glut conditions appeared in the world oil market in 1981, partly through economic recession in principal oil-importing countries and partly through the delayed reaction of oil users to the revaluation of this source of energy achieved by OPEC in 1973–4. Nigerian oil production in 1981 and the first half of 1982 was 40 per cent down on the level attained in 1979, and some reduction of the export price was made after August 1981. A substantial current deficit must have appeared in the balance of payments in 1981, and the foreign exchange reserves were being run down rapidly from the middle of that year. Failing any sustained recovery in oil sales, administrative measures to restrain imports were reintroduced and intensified early in 1982.

In short, the Nigerian balance of payments in the 1970s and after was dominated by the value of oil exports. On the one hand, oil vastly enhanced purchasing power over imports (specifically, the government's purchasing power). On the other hand, political events affecting oil supplies, and economic adjustments affecting the demand for oil, made export prices or saleable output extremely volatile. The government reacted to this volatility partly by variations in its own spending (making rapid, and potentially wasteful, increases in investment spending in 1975 and 1980 in particular) and partly through administrative controls on imports (strengthening restrictions in 1979 and 1982 in particular). External borrowing also provided room for manoeuvre in face of adversity. It was resorted to quite heavily from 1978 (some $4.6 billion being raised in the Eurocurrency market by the middle of 1981); even so, the debt service ratio (ratio of payments on external debt to value of exports) had reached only some 5 per cent by 1981.

Table 21 shows the evolution of the Ivorian balance of payments from 1970 to 1978. As in Nigeria, there was a pronounced upward trend in the value of exports; it was most rapid in 1976–7, when cocoa and coffee prices rose briskly. The trade balance was positive throughout the period, notwithstanding the country's dependence on imports for its oil supplies; but these surpluses were more than offset by deficits in non-factor services (freight, insurance, travel, management and agency fees and commissions), investment income, and the unrequited transfers which are mainly workers' remittances. In consequence, there were regular current deficits. Between 1971 and 1975, the balance of payments was also in deficit overall, and external reserves were run down to an extremely low level by 1975. The avoidance of further overall deficits in 1976–8 required increases in official borrowing abroad, especially in the last year of the period.

While export proceeds increased fivefold (in current dollars) between 1970–1 and 1977–8, the increase in the sum of the negative invisible items was eightfold. The weight of these invisible items in the balance of payments, and their tendency to grow in relative importance, are partly attributable to the extensive employment provided in the Ivory Coast to foreign labour, enterprise and direct investment. In addition, official debt servicing has become of increasing significance since the public sector has assumed the major share in capital formation but has been unable to provide from its revenues more than a fraction (35 per cent in 1975) of its net financing requirements.[23] This official external debt has several familiar characteristics. Much of it was contracted by parastatals and quasi-autonomous public bodies without benefit of centralized control. It contained large elements of suppliers' credits and bank borrowing. The uses to which it was put included investments, notably in education and public housing, yielding returns that were

Table 21 Ivory Coast: balance of external payments, 1970–8 (US $m)

	1970	*1971*	*1972*	*1973*	*1974*	*1975*	*1976*	*1977*	*1978*
Exports (fob)	496	495	596	862	1,253	1,239	1,735	2,412	2,616
LESS imports (fob)	432	399	460	701	894	1,012	1,160	1,596	2,043
Balance of trade	64	96	136	161	359	227	575	816	573
Services and income (net)	−86	−175	−204	−311	−327	−469	−565	−692	−996
Unrequited transfers (net)	−12	−27	−50	−69	−93	−142	−259	−300	−418
Current balance	−34	−106	−118	−219	−61	−384	−249	−176	−841
Private capital transactions (net)	33	21	32	128	119	153	151	177	333
Official capital transactions (net)	39	60	4	73	50	53	121	205	588
Errors and omissions	−7	9	–	7	−21	14	10	−24	8
Overall balance	31	−16	−81	−10	−87	−164	33	182	89

Note: CFA franc totals have been converted to US dollars at annual average rates of exchange.

Source: BCEAO, *Notes d'information et statistiques*.

very long-run, indirect, and probably low.[24]

The Ivorian debt service ratio rose from less than 7 per cent in 1970 to nearly 11 per cent in 1975. The World Bank Mission that visited the country in 1975 reported that a ratio of 15 per cent would have to be considered high in so open an economy.[25] That ratio was reached in 1979 as a result of major expansion of public investment, under the stimulus of the export boom of 1976–7, and of the fall in the Ivorian terms of trade that began in 1979. In the same year, the ratio of external debt to the GNP estimate reached 40 per cent.

A measure of the pressure of external payments felt by the Ivorian government is its reaction to falling cocoa export prices in 1979. In an effort to arrest or reverse this decline, Ivorian cocoa was withdrawn from the world market in October of that year. The experiment failed: in June 1980, the government was forced to allow resumption of sales at prices that were now, in fact, even lower. Subsequently, the government refused to be a party to the new International Cocoa Agreement, made in mid-1981, on the ground that the range within which it was hoped to stabilize market prices had been set too low. Ivorian diplomacy continued to make much of the unfairness of the prices paid by importers to producers of primary products, a concern that was as much fiscal as it was solicitous for Ivorian farmers.

As another reaction, the reorganization of the parastatal sector, which had been tentatively begun in 1975, was pushed ahead with more vigour. According to President Houphouët-Boigny in June 1980: 'these state companies have been largely responsible for the catastrophic situation of the state finances.'[26] In 1981, a line of credit over three years was negotiated from the IMF, in return for commitments to limit other foreign borrowing and domestic credit, to reduce the subsidization of state enterprises, and to improve efficiency in budgetary procedures.

As in Ghana in the 1960s, so in the Ivory Coast in the 1970s, official borrowing from abroad at commercial rates and with insufficient care in the choices of investments was instrumental in creating payments difficulties. In both cases, a fall in the world market value of cocoa, and hence in public receipts of foreign exchange, was less the cause than a precipitant of crisis in the balance of payments – an exposure of the mistakes that had been made in investing borrowed funds. In the Ivorian case, the start of crude petroleum production in 1980, and the likelihood that substantial public revenue will be forthcoming from this source by about 1985, suggest that the payments difficulties will not become chronic (as they have in Ghana) and that the government will recover its external credit.

While neither Nigeria nor the Ivory Coast found it easy to manage external payments in the 1970s, the smaller West African countries experienced more acute problems. By 1979, the burden of external debt was about as heavy in Liberia, Senegal and Sierra Leone as in the Ivory Coast, and it was even heavier in Guinea, Mauritania and Togo.[27] Other countries, like Ghana, avoided heavy external indebtedness at this time, not because they did not need to borrow but because they had long ceased to be creditworthy. The oil price rises of 1973–4, and the contemporaneous increases in the costs of imported cereals, appear not to have been decisive in producing these debt burdens and payments problems, since the Sahelian drought occasioned large inflows of foreign aid and the prices of some West African exports were also buoyant at this time;[28] the commodity terms of trade appear actually to have improved for Benin, Ghana, Senegal and Togo between 1972 and 1974.[29] Much more damaging was the second surge in oil prices that began in 1979, since it coincided with stationary or declining prices for most other West African exports.

It would nevertheless be misleading to represent the governments of the smaller countries as the helpless victims of forces over which they had no control. Their troubles have been at least as much of their own making. In Ghana, for instance, a system of political economy has been created since the early 1960s in which exporting is made unrewarding and importing highly profitable by an extreme overvaluation of the currency, and in which administrative attempts to counteract this distortion increasingly deflect economic activity from lawful and taxable into unlawful and untaxable channels. The consequences have been such as to threaten the territorial integrity of the state. Mali has been another conspicuous example of the damaging effects of price distortions wrought by public policy.

In Sierra Leone, payments difficulties since 1976 reflect an inability to restrain public expenditure either within the means available or the targets projected, and regular recourse to deficit financing. In Liberia, a downturn in the income terms of trade from 1974 was accompanied by a marked expansion in public investment and a proliferation of public corporations, with the result that the public debt had more than trebled (from $203 million to $667 million) by 1979. At the centre of Mauritania's payments difficulties lay the government's involvement from 1976 to 1979 in the war in the Western Sahara,[30] and its undertaking of several large infrastructural and industrial projects on the strength of foreign suppliers' credits, while a similar phase of large-scale and externally financed public investment was launched in Togo on the miscalculation that the extremely high phosphate prices of 1974–5 would persist; see the huge investment ratios estimated for these countries in Tables 5 and 6. By the beginning of the 1980s, the governments of Sierra Leone, Liberia, Mauritania and Togo had all to seek the agreement of their foreign creditors for debt reschedulings and the support of the IMF for measures intended to stabilize their finances. The problems of these governments had originated in the mid-1970s, between rather than during the phases of increase in imported fuel costs, and, like the payments problems experienced by some West African countries in the 1960s, they are chiefly to be attributed to financial irresponsibility.

Exchange rates

The external values officially put on West African currencies diverge from market reality in proportion to the stringency with which the government controls external transactions (or, at least, lawful transactions) and the vigour with which it has inflated domestic purchasing power. At one extreme lies Guinea, whose currency, the syli, almost since its inception in 1960, has been 'effectively worth nothing outside Guinea and very little inside Guinea'.[31] At the other extreme is Liberia, whose paper money is not merely freely convertible into, but actually is, the US dollar.

Next to Guinea, Ghana has the distinction of possessing the most valueless currency in the region. The Ghanaian pound was originally issued at par with sterling. No alteration in the official exchange rate was involved when this currency was decimalized (as cedis) in 1965, although the rate was already overvalued. Following the fall of the Nkrumah government, the cedi was devalued by 30 per cent in July 1967 as a step toward liberalization of the import regime. In November 1971, in face of rapidly increasing imports and a temporary drop in cocoa earnings, the cedi was pegged to the depreciating dollar, and during the next three months it was devalued against the dollar by a net 20 per cent.[32] Import and exchange controls were then reimposed. A succession of large budget deficits from 1974 produced rapid inflation and another devaluation, of 58 per cent against the dollar, in August 1978. The official value of the cedi had then fallen to $0.36 from $1.40 in 1966, or by almost three-quarters. Even so, this official value remained a large multiple of the cedi's black-market value after 1978.

The Nigerian pound was also introduced at par with sterling, but it was allowed to appreciate against that currency when sterling was devalued in November 1967, the par value being maintained instead in terms of gold and, for a time, US dollars. The dollar value of the Nigerian pound rose from $2.80 to $3.04 with the dollar devaluation in December 1971, and the value of the naira was consequently fixed at $1.52 when that currency unit replaced the pound, at the rate of two for one, at the beginning of 1973. That exchange rate was maintained despite the second dollar devaluation in February 1973, but from April 1974 Nigeria followed an independent exchange rate policy under which the naira was appreciated against both the dollar and sterling, reaching a peak of over $1.87 at the end of 1980. The policy option of devaluing was not taken when the need to reduce imports was felt in 1978 and 1982, despite clear indications that the naira had become substantially overvalued.

In Sierra Leone, a national currency unit, the leone, was introduced in 1964 at the rate of two for each West African Currency Board pound and the same par value in sterling. The leone was devalued along with sterling in November 1967, and the exchange rate was maintained at two leones to the pound when sterling was floated in 1972. In November 1978, the strengthening of sterling led to a decision to align the leone with the SDR;[33] the effect was to devalue the leone against sterling by about 5 per cent. Subsequent pressure by the IMF in favour of further devaluation was resisted by the government.

The Gambia's currency has always been pegged to sterling – both the Currency Board pound and the dalasi, which replaced it at the rate of five for one in 1971. In March 1973, the parity with sterling was raised from five to four dalasis in order to arrest increases in the dalasi prices of imports from non-sterling sources.

The CFA franc used in common by the member-countries of the BCEAO (Benin, the Ivory Coast, Niger, Senegal, Togo, Upper Volta and – until 1973 – Mauritania) is pegged to the French franc at the rate of fifty to one. In Mali, after five years of monetary autonomy, the national currency was devalued by 50 per cent in May 1967 and pegged to the French franc at the rate of one hundred to one. There is also a fixed rate of two Malian francs per CFA franc, but Mali has not been able to negotiate its re-entry to the UMOA.

Since the convertibility of these African francs into French francs is guaranteed by the French Treasury and both current and capital transfers are mostly unimpeded within the franc zone, the official values have not become seriously detached from market values. They have moved with the external value of the French franc, in the same way as the leone followed sterling until 1978, and they have therefore fallen against non-franc currencies on the occasion of French devaluations (August 1969, September 1981, June 1982). These involuntary exchange rate adjustments are a price paid for France's underwriting of the external value of the African francs. But it is disingenuous to represent them as functionless from the point of view of the African countries, and evidence only of a neocolonialist relationship of subordination to French interests.[34] Even in the absence of the French guarantee of convertibility, some compensating action in the event of a French devaluation might well be required of the African countries, given the importance of France as a trading partner, in order to maintain their ability to compete at home with imports and the competitiveness of their exports to France.

Elsewhere in West Africa, deficit financing of governments and administrative controls on external transactions have produced overvaluations of currencies at the official rates – extremely so in Ghana and Guinea, but also appreciably so in Nigeria since the civil war and in Sierra Leone since 1975. The effects are artificially to cheapen imports and to deter any exports in which home costs are a substantial proportion of market value,[35] and thus to aggravate the difficulty of maintaining external balance. It is undoubtedly true that these countries cannot fully enjoy the advantages offered by international trade so long as their currencies remain overvalued. On the other hand, removal of an overvaluation is practically difficult because political interests become entrenched in cheap imports, as they do in any public subsidy. This political obstacle mainly explains the resistance of the governments concerned to suggestions that they should devalue. But it may be added that they could also be right in believing that a devaluation would be economically fruitless. So long as the devaluation was not accompanied by monetary discipline, and in particular avoidance by the government of inflationary financing of the public sector, the removal of price distortions would be short-lived, if it became apparent at all; this much has been demonstrated by the successive devaluations of the Ghanaian cedi.

International economic integration

As in many other colonial territories, external trade and transactions in West Africa were largely free from official control until the 1930s. Tariffs were imposed only for revenue-raising purposes, and the direction of trade was not formally regulated. Not only was an 'Open Door' for commercial intercourse widely regarded as an aspect of colonial trusteeship, but also Great Britain and France had formally agreed in 1898 not to discriminate against each other in most of West Africa, and the immunity had been extended to other countries through most-favoured-nation clauses in other treaties. This liberal order was first breached by reason of the world depression. Sierra Leone and The Gambia were brought into the system of British imperial preference created by the Ottawa Conference of 1932. From 1934, quotas were applied to Japanese goods imported into British West Africa. By 1936, all of AOF had been brought within the preferential tariff regime of the French empire. About the same time, the first attempts were made to create the fiscal means of stabilizing export-crop prices in the French territories.[36]

These controls were much intensified during the Second World War and in its immediate aftermath. Both statutory exporting monopolies and the administrative licensing of imports were introduced in the British territories soon after the outbreak of war. Their effect was to produce unrequited exports, or enforced loans to the UK government. These institutions survived the war, partly in order to protect the British balance of payments during the period of dollar shortage, but it was found possible to relax the severity of the import licensing in the later 1950s, immediately before the independence of Ghana and Nigeria. The incorporation of the British territories into the sterling area had been paralleled by that of AOF into the franc zone. In the latter case, the bilateral relationships were stronger and more durable, since they included the payment by France of a *surprix* (a price higher than the world market price) for substantial quantities of West African exports as well as the toleration by the African countries of relatively expensive imports from France. The decisive event in ending this preferential connection was less the independence of the African territories than the creation of the European Economic Community.

After independence, the marketing boards, with their monopolistic powers over export crops, were retained in the former British territories, as were the *caisses de stabilisation* which had been instituted in the French territories in the mid-1950s and which were to become, like the marketing boards, mainly fiscal agencies. In all countries, tariffs were used to protect local industries as well as to raise revenue, continuing a practice that had appeared in the 1950s. Import licensing was more variously used. At the extreme, it was employed in a thoroughgoing way (or as thoroughgoing as the public administration could manage), not only to regulate the total of foreign exchange expenditures, but also to select among imports, sources of imports, and importers. External transactions in payment of services and capital transfers were similarly controlled. Guinea entered such a regime immediately after its independence in 1958. Comprehensive foreign exchange controls have also operated in Ghana since 1961, save for a liberal interlude in 1969–71. Nigeria moved away from administrative controls of external transactions until the civil war in 1967; subsequently, it has relaxed or intensified these controls mainly in reaction to shifts in the balance of current payments. In Sierra Leone, the view was long held that external transactions were not, in practice, amenable to official control, but in 1975 the authorities determined nevertheless to attempt such control. The member-countries of the BCEAO have eschewed administrative regulation of their transactions with the franc zone, the European Community, and one another. Liberal import regimes have also been maintained in Liberia and The Gambia.

The causes and consequences of export monopolization, and of import and exchange controls, will be examined in Chapters 5 and 6. The rest of the present chapter will be concerned with the association of West African countries with the European Community, and with their attempts to coordinate economic activities among themselves through preferential trading arrangements within the West African region.

Association with the EEC and the creation of a West African customs union might well be seen as alternative directions of institutional change, when independence presented formally the opportunity to escape such elements of enforced bilateralism with

Great Britain or France as had survived from the late colonial period. On the one hand, a wider integration of West African activities with the economically advanced countries of Western Europe might be sought; on the other hand, this transoceanic integration might be resisted in favour of the development of complementary economic activities within Africa. If individual West African countries were economically too small for such a development to be feasible on a national basis, it might be attempted instead on a continental, or at least a regional, basis. International integration within Africa might therefore be opposed (and, in the 1960s, was opposed)[37] to the international integration of Africa with countries overseas. But since both kinds of integration have been sought by West African governments, it would seem that these alternatives have been found, in practice, not to be mutually exclusive.

The arguments for and against African customs unions have become the arguments for and against protection of African industries against imports. To understand how this has happened, it is convenient to begin with Viner's criteria for judging the desirability of a customs union.[38] Viner distinguished between the 'trade creating' and the 'trade diverting' effects of a union. Trade creation meant an increase in trade among the member-countries of the union, consequent on the removal of trade barriers. To the extent that it occurred, high-cost production in some parts of the union was being displaced by low-cost production in other parts, and resources in the union were more efficiently employed. Trade diversion, on the other hand, was what happened when, because of customs preferences, greater exports from one member of the union displaced in another member, not some of the latter's domestic production, but some of its imports from the rest of the world. To the extent that trade diversion occurred, high-cost production within the union was displacing low-cost production outside the union, and income within the union would tend to be reduced.

The balance between trade creation and trade diversion – not simply the amounts of trade created and diverted, but these amounts weighted according to the differences in costs of production from different sources, and taking into account variations in costs with the scale of production – would determine whether or not the union was advantageous to its participants.

Trade creation would appear likely to predominate in a union where the member-countries have low ratios of external trade to GDP, but where a high proportion of this external trade is among themselves. The reasons are that, the more self-sufficient a country, the more its union with others is likely to displace domestic production by imports; and the greater the concentration of the existing external trade among the member-countries, the less likely is trade diversion from the rest of the world.

Judged by these considerations, the prospects for a West African customs union are highly unpromising. As was noticed in Chapter 1, export ratios in the region are substantial (about 23 per cent of GDP for West Africa as a whole in 1979), and only a small proportion of external trade (3 per cent of recorded exports in 1979) takes place within the region. A West African customs union therefore violates both of the *a priori* conditions for trade creation's outweighing trade diversion. Having regard for the composition of West African exports and imports and the present productive capacities of the region, it might be expected that such a union would have, in practice, little effect on the allocation of resources, but, to the extent that it was effective, the results would appear likely to be disadvantageous to the participants.

The analysis so far does not explain how customs unions arrive on the policy agenda. If trade creation represents gain, and trade diversion loss, every government would seemingly be best advised to seek universal free trade, rather than preferential trading groups. Free trade would bring all the attainable benefits of trade creation without any offsetting loss through trade diversion. Yet in practice all governments impede external trade, to a greater or lesser extent, thus apparently sacrificing domestic income. Then the question arises whether a customs union will reduce this loss (if, on balance, the union is trade-creating) or increase it (if, on balance, the union is trade-diverting). Why are the trade barriers there in the first place, and why should they be retained against non-union members in face of the apparently unalloyed gains that might be won from free trade?

Freedom of trade is impeded for various reasons, including the raising of revenue, protecting the balance of payments against the consequences of monetary autonomy, and assuaging special economic interests. In addition, there are economic arguments that the protection of domestic activities against imports can promote economic growth.

Economic growth in any country is causally related to changes in the quantities and qualities of productive resources, in forms of economic organization, in technology, and in consumers' tastes. Free trade would maximize income in conditions where these changes were independent of the composition of domestic output. If, on the other hand, it is accepted that these changes depend, at least in part, on the kinds of economic activity being carried on in the country, a case emerges for departing from free trade in order to help reshape the composition of output and thus, perhaps, to promote economic growth.

Economic advocates of protection in areas like West Africa have therefore postulated negative effects on economic growth of unprotected economic structures (underemployment of resources, instability, poor long-run prospects for export sales) and gains potential in the transformation of those structures under cover of protection (encouragement of capital inflows, development of skills and technical knowledge, increases in the versatility of resources).

For many of the new activities (chiefly identified in manufacturing) established under cover of protection, unit-costs vary with the scale of production and there are sizes of plant below which costs become inordinately high. Since West African countries are mostly small in purchasing power and markets are therefore narrow, it follows that, in many cases, these deliberately induced structural transformations are either quite limited in practice or cripplingly expensive. The case for a West African customs union then becomes that, by widening the markets available to protected activities, the union will permit more transformation, or less costly transformation, than is possible so long as the countries of the region maintain trade barriers against one another. Implicit in this conception of a union is the principle that the protected manufacturing plants would be non-competing, except in activities where the market was large enough to permit more than one plant of a tolerably efficient size.

Such a union is valued, not because it may, on balance, result in trade creation but, on the contrary, because it will allow more trade diversion, or cheaper trade diversion. This reversal results simply from preferring protection to free trade. The union is no longer an incomplete movement toward free trade, but instead a means of increasing the efficacy of protection. To criticize such a union on the ground that it is trade-diverting is beside the point, for it is intended to divert trade. The arguments for (or against) the union are essentially the arguments for (or against) protection.

West African governments have often wished to protect new industries in their territories, and they have looked kindly, at least in principle, on proposals for regional customs unions. At the same time, they have sought free or preferential access for their exports to the markets of their trading partners in Western Europe. To a substantial extent, these apparently inconsistent objectives have proved reconcilable. The reasons lie in the nature of West African exports, which are mostly non-competitive with European production, and the readiness of the West European governments to allow breaches in, and finally abrogation of, the principle of reciprocity in their trade agreements with Africa.

The association of African countries with the European Economic Community resulted from a French initiative in 1956, at a late stage in the negotiation of the Treaty of Rome. Part IV of the Treaty allowed for the creation of free trade areas[39] between the EEC members as a group and each of their eighteen African dependencies. A proviso was made that the associated territories might retain against the EEC such import duties as were required for raising revenue or for the protection of 'development and industrialization'. A new multilateral aid institution, the European Development Fund (EDF), was to be created by the EEC members for the benefit of the associates. Grants totalling $581 million were made from this fund during the period 1958–62, mostly to the French (or formerly French) associates. When the associates became independent states, only one – Guinea – repudiated association; Dahomey, the Ivory Coast, Mali, Mauritania, Niger, Senegal, Togo and Upper Volta remained West African associates of the EEC.

Outside the EEC and the associated territories, Part IV of the Treaty of Rome initially attracted much hostility. It was widely believed that association was an attempt by France, or by the EEC members collectively, to retain some kind of dominion over the associated territories after their independence was granted, and to obstruct a possibile unification of African states. The legality of association was questioned, chiefly by Nigeria, on the grounds that the rules of the General Agreement on Tariffs and Trade[40] prohibited the creation of new preferences in international trade, except as a result of the formation of a customs union or free trade area, and that the free trade areas produced by association were not genuine because of the proviso allowing the associates to maintain duties against the EEC. Opposition was mounted in countries (like Nigeria and Ghana) whose own exports would, it was believed, be prejudiced by the duty-free access of the associates' exports to the EEC market. This disadvantage was greatly exaggerated, at least so far as products currently exported were concerned. The EEC in fact had zero tariffs on minerals and many raw materials. For crops including cocoa, coffee and palm oil, the effects of the rather low EEC tariffs against non-associated supplies were to hold up prices in the EEC market (since the associates could not, in practice, supply all the requirements of that market), and thus to confer on the associated suppliers small quasi-rents at the expense of EEC consumers. The non-associated suppliers would be disadvantaged only in the unlikely eventuality that these small quasi-rents encouraged such an expansion of output of the crops in the associated countries as to depress the prices of the crops in the world and EEC markets.

In the event, EEC imports from the associated states rose less swiftly during the period 1958–62 than imports from non-associated developing countries. The former French dependencies continued to export mainly to France, where they still enjoyed the premium prices arranged through colonial bilateralism, and the EEC countries other than France were mainly supplied from non-associated sources.

A renewal of the association arrangements, known as the Yaoundé Convention, was negotiated in 1962 and, after delay occasioned by the failure of the British attempt to join the EEC, it became operative from 1964 to 1969. In response to international pressures, the EEC tariffs on supplies from non-associated developing countries were reduced, and association was made available in principle to any non-associated state in geographical regions which were economically of the same structure as that of the original associates – a concept which certainly embraced all the non-associated states in tropical Africa. A new European Development Fund was constituted. $730 million was to be made available, mainly as grants, over five years. Of this total, $230 million was earmarked as aid for agricultural efficiency and diversification, mostly for the benefit of former French dependencies which would have to adjust to EEC or world prices for their exports as France eliminated its duties and quantitative restrictions on imports from the non-franc associated states. The phasing out of the French *surprix* arrangements was in fact completed by the beginning of 1965 for coffee, cotton and palm produce, and by the beginning of 1968 for groundnuts.

Only one West African state, Nigeria, took up the invitation extended at Yaoundé to eligible non-associates[41] – choosing not to adhere to the Convention but to negotiate with the EEC an association *sui generis* that was limited to reciprocal tariff concessions. The Lagos Agreement of 1966 was hailed as a great achievement of Nigerian diplomacy, but its ratification was delayed, and finally prevented, by discords arising from the Nigerian civil war, and it never became operative. Its economic effects on Nigeria were limited to the costs of the salaries and expenses of the civil servants who negotiated the agreement over a period of about three years.[42]

After 1964, as before that date, imports into the EEC rose faster from the non-associated than from the associated developing countries. Spokesmen for the associates therefore pressed for additional subsidization by European consumers when negotiations for a new Convention began in 1968. (Since the fall of Nkrumah in Ghana in 1966, less had been heard of European interference in African affairs, and more of inadequate European support for Africa.) It was not feasible to raise the rates of preference enjoyed by the associates, since opinion in the EEC (at least, among the Germans and Dutch) was strongly in favour of removing or reducing those preferences. Instead, the EEC members were urged to agree to make purchases of quotas of the associates' export products at guaranteed minimum prices. This proposal seems to have envisaged the revival, on a West European scale, of the French *surprix* arrangements which had just disappeared. In the event, very little was achieved in this direction under the second Yaoundé Convention signed in 1969, but the influence of the *surprix* exemplar on official economic thinking in francophone Africa has persisted, and underlies support of proposals made in the 1970s for the construction of a New International Economic Order.

The total provided the African associates by the third EDF was $918 million over a period of five years and seven months ending in January 1975. The new Convention further reduced some of the preferences given the associates in the EEC market; also, it provided that free trade between the associates and the EEC should not obstruct adoption in the EEC of the Generalized System of Preferences[43] for the exports of developing countries. The associates too were given some room for manoeuvre; the Convention committed the EEC members to waive their most-favoured-nation rights in the associated states, if this action would further African regional groupings.

In 1975, the Yaoundé Convention associating nineteen African states with the EEC was replaced by the Lomé Convention establishing special economic relationships between the EEC and forty-six so-called ACP (African, Caribbean and Pacific) countries, which included all fifteen of the mainland West African countries.[44] Primarily, this change resulted from the entry to the European Community of the UK (along with Denmark and Ireland) at the beginning of 1973, and a recognized need to protect former British dependencies against the discrimination by which they were now threatened in the UK market. Negotiations to continue association under the Yaoundé Convention were therefore absorbed in negotiations to safeguard the interests of these Commonwealth countries. On the ACP side, the leading role was played by the federal military government of Nigeria, whose hand was strengthened by the circumstance that it had little direct interest in the outcome of the negotiations and was mainly concerned to increase its standing and influence in and beyond Africa.

By the Lomé Convention, imports of ACP products into EEC countries were almost entirely freed from customs duties. About three-quarters of these imports by value were already duty-free, including manufactures which had free access under the Generalized System of Preferences. But small preferences in favour of the ACP states were implied for agricultural products including cocoa, coffee, palm oil and pineapples; and there were larger preferences for ACP manufactures against supplies other than those from other developing countries and from the EEC countries themselves. In order to qualify as ACP products, manufactures were to consist as to at least one-half in value added in the ACP countries considered as a single customs territory. The free access of ACP supplies was qualified by provisions safeguarding European producers; it could be limited in connection with implementation of the Common Agricultural Policy[45] of the EEC, or if it led to 'serious disturbances' in an economic sector of an EEC country. These safeguards were not immediately of much relevance to West African exports.

The ACP states were not required to give in return free access to any EEC exports. They undertook not to discriminate *against* the EEC; but even this requirement could be set aside in order to encourage trade among ACP states themselves, or between ACP states and other developing countries. The principle of reciprocity, which had been present, at least symbolically, in the association agreements ever since 1958, was thus abandoned. One reason was that the UK had received few formal preferences in the markets of the former British dependencies, and there was no enthusiasm among the latter for creating preferences afresh and losing customs revenue as a result. Another was that non-reciprocity by developing countries was embodied in the Generalized System of Preferences. A third was that the USA strongly objected to preferences for European and against American goods in the markets of developing countries.

Multilateral EEC aid to the ACP states was to total 3,390 million EUA[46] (worth, in 1975, about $3,160 million), of which 3,000 million would constitute the fourth European Development Fund and the remainder would be lent by a European Investment Bank. Over two-thirds of the Fund would be distributed as grants. An innovation was the earmarking of 375 million EUA for stabilization of ACP export earnings from specified products under the scheme that became known as Stabex.

Twelve product groups were originally specified under the Stabex scheme; they included groundnuts, cocoa, coffee, cotton, palm produce, timber, bananas and iron ore. If the earnings of an ACP country in any year from exporting such a product to the EEC fell below the previous year's earnings by more than a predetermined proportion (the 'trigger threshold'), and the earnings from this product accounted for more than a predetermined proportion (the 'dependence threshold') of the country's total earnings from exports to the EEC on the average of the previous four years, the government concerned could request financial compensation to make up the shortfall. The earnings were to be measured in current values, uncorrected for changes in import prices. For twelve of the ACP countries (including Ghana, the Ivory Coast, Liberia, Nigeria,[47] Senegal and Sierra Leone), the trigger and dependence thresholds were both 7.5 per cent. For the other thirty-four (known as the least developed, landlocked or island states, and including the nine other West African countries), the thresholds were both 2.5 per cent. For the poorest countries (defined to include Benin, The Gambia, Guinea, Guinea-Bissau, Mali, Mauritania, Niger, Togo and Upper Volta), the Stabex transfers would be grants; for the others, these transfers would normally be interest-free loans over five years.

West African governments received just over one-half of the total transfers from the first Stabex fund. Major beneficiaries were Senegal (65 million EUA for groundnuts), Mauritania (37 million EUA for iron ore), Niger (22.5 million EUA for groundnuts), Benin (15.4 million EUA for cotton), and the Ivory Coast (15 million EUA for timber).[48] The scheme was arbitrary, and even eccentric, in its operation; for example, minerals other than iron ore were excluded, and the Ivory Coast received compensation for timber at a time when its other exports were rising rapidly in value. It was also extremely attractive to the recipients. Stabex funding came as grants to the poorest of them. It was a form of aid over whose use all recipients had almost complete discretion. Because of that characteristic, it could be paid over quickly, in sharp contrast to the rate of disbursement of the rest of the EDF.

In October 1979, a second Convention was signed at Lomé, renewing the special economic relationships between the EEC and the ACP states for five years beginning March 1980. There were now fifty-eight ACP states; the additional twelve were mostly small island states, among them Cape Verde. Little change was made in the trading arrangements instituted in 1975. Multilateral aid from the EEC was fixed at 5,227 million EUA (worth, in 1980, about $8,025 million) over the five years, including 4,542 million EUA in a fifth European Development Fund.[49] The funding earmarked for Stabex was 550 million EUA. The trigger and dependence thresholds were reduced from 7.5 to 6.5 per cent for the better-off ACP states, and from 2.5 to 2 per cent for the poorer, and those recipients required to repay Stabex transfers were now to be allowed a grace period of two years. The number of products or product groups covered by the scheme had been increased to twenty-nine; of most significance for West Africa among the additions was rubber.

The first Stabex fund had been sufficient to meet in full all the claims made on it. Under the second Convention, allowable claims for 1980 were about double the funding available[50] and had to be scaled down accordingly. Just over one-half of the transfers made for 1980 went to West African states, principally Senegal and The Gambia for groundnuts and the Ivory Coast for coffee. The terms of trade of many ACP states continued to deteriorate in 1981, and there was an even greater disproportion between the allowable claims for that year and the ability to meet them from the Stabex fund. Pressure greatly to increase the sums available to Stabex was resisted; the commitment of the EEC members to stabilizing the revenues of the ACP governments was clearly not unlimited.

The chief novelty introduced in 1979 was Minex or Sysmin, a scheme resembling Stabex but with somewhat different purposes. Minex referred to five groups of mineral exports – copper (including associated production of cobalt), phosphates, manganese, bauxite and alumina, and tin; in addition, iron ore was to be moved from Stabex to Minex in 1984. Under this scheme, the trigger threshold would be a 10 per cent fall, not in export earnings, but in 'production or export capacity'. The dependence threshold would be 10 per cent of export earnings for the least developed, landlocked and island ACP states, and 15 per cent for the others. (In West Africa, Guinea for bauxite and alumina and Togo for phosphates would therefore be eligible as well as Liberia and Mauritania for iron ore.) Transfers from the 280 million EUA allocated to Minex would take the form of long-term, low-interest loans. They would not be used (like the Stabex transfers) for the general support of government revenues, but for investments in projects intended to arrest the decline in production or export of the affected mineral. Thus, the Minex scheme was concerned with obtaining continuity in mineral supplies to Europe, as well as or rather than with stabilizing African economies; in fact, it originated in West German reaction to the effect on copper supplies of the so-called Shaba wars in Zaïre in 1977–8.

The effects of the trading relationships maintained since 1958 between the EEC and West African countries have almost certainly been trade-creating for the latter, since those relationships replaced (especially in the francophone countries) narrower preferential arrangements with the former metropolitan power. The effects in Western Europe have possibly been trade-diverting at the expense of developing countries other than the associated or ACP states, but not to an important degree; the ACP states supply only a modest proportion of EEC imports (5.6 per cent in 1982), and the preferences in their favour are neither general nor very large. Whether the effects in Western Europe can be trade-creating in the future will depend on restraint in the use made of the provisions in the Lomé Convention safeguarding European producer interests; exports of cotton textiles from the Ivory Coast are of most immediate relevance in this connection.[51]

West African governments have perhaps valued their relationship with the EEC more for its aid-diverting effects than for its effects on their countries' trade. Bilateral aid by the EEC members has always overshadowed aid delivered through the Community's institutions. The creation of successive European Development Funds has not necessarily increased the total amount of aid given. But the proportion of West European aid (especially of West German and Dutch aid) given to the associated or ACP states has probably been increased because this multilateral aid agency has existed with a particular geographical remit. The francophone West African governments obtained the lion's share of the first three EDFs. Even following the enlargement of the EEC and the creation of the ACP grouping, they received 35 per cent of the disbursements made by the end of 1980 from the fourth EDF.[52]

It has already been said that the preferential trading arrangements between the EEC and West African countries have been formally compatible with attempts economically to integrate the latter by customs unions. Such a West African union was, in fact, formed as early as 1959 – the *Union douanière des états de l'Afrique occidentale*, consisting of Dahomey, the Ivory Coast, Mali, Mauritania, Niger, Senegal and Upper Volta. Like the common currency and central bank of those countries, their customs union was a continuation of colonial institutions. Unlike the monetary arrangements, it proved to be untenable after independence. In particular, the Ivorian authorities refused to leave their industries and the transit trade with interior countries unprotected against Senegalese competition. By the mid-1960s, the union had become 'a purely nominal association'.[53] In 1973, it was reconstituted (without Dahomey or Benin) on a firmer basis as the *Communauté économique de l'Afrique de l'ouest* (CEAO).

Several other organizations of West African states were to appear in the early years of independence, but their purposes were either primarily political or more narrowly functional than those of a customs union.[54] A proposal for a customs union embracing all West Africa originated with the identification of the region by the United Nations Economic Commission for Africa as an area of potential economic cooperation, initially with reference to the establishment of an iron and steel industry. In 1967, all fourteen of the independent states of the region signed articles of association for the formation of a West African Economic Community (WAEC).

The agreement on WAEC would have remained an empty declaration of intent, like many other manifestations of African unity, but for the conviction of the Nigerian military government, when the civil war ended, that the creation of such a community was required for the security of Nigeria, in the long-run economic interests of the country, and above all as a means of establishing Nigerian claims to political leadership in Africa. This development of Nigerian ambitions coincided with, and was of course encouraged by, rising government revenues from petroleum. Between 1970 and 1975, the Nigerians campaigned to influence the thinking of their neighbours on the merits of WAEC, freely spraying largesse in the direction of those who might be inclined to heed French (or Ivorian, or Senegalese) warnings against Nigerian ascendancy in West Africa.[55] This campaign could not prevent the formation of CEAO, the francophone union, but the six signatories of that agreement were among the fifteen who subscribed in May 1975 to the Treaty of Lagos establishing the Economic Community of West African States (ECOWAS).

The ultimate objective of the Treaty of Lagos was to create an economic union in the West African region. Not only were duties and other restrictions on trade among the members to be eliminated and a common external tariff constructed, but also there was eventually to be free movement of persons within the Community, national policies relating to agriculture, industry and transport were to be harmonized, and the disparities in levels of development among the members were to be removed. These far-reaching objectives imply a vast amount of detailed negotiation among the governments of the region concerning the timing of implementation, the resolution of technical difficulties, possible waivers and exemptions, the location of new industries based on the common market, and compensation for transitory damages such as loss of customs revenue. Even within a country, agreement on some of these matters (fiscal redistribution, industrial location, labour mobility) can be extremely taxing, as experience in Nigeria itself has shown.[56] In a region composed of sixteen national states, among which there are important cleavages in language, institutions and administrative traditions, the task of harmonizing interests is even more daunting, and the outcome highly problematic. The difficulties of achieving ECOWAS are compounded by the concurrent pursuit of the liberalization programme of the CEAO, which not only has a different timetable of implementation but also aims at preferential trade, rather than free trade, in the products of union members.[57]

The Treaty of Lagos envisaged regional free trade in West

African products by 1987 and erection of a common external tariff by 1993. This timetable has slipped. The trade in unprocessed goods and handicrafts became free in 1981, at least in principle, but the programme for liberalizing the trade in manufactures, reported in June 1982 to be 'trembling on the brink of becoming operational',[58] was then not expected to be completed until 1989. Difficulty has also arisen, especially in Nigeria, over the provision of the treaty promising free movement of persons, even though all that has been achieved so far is the waiving of visas for visits not exceeding 90 days.[59]

The trade among West African countries in foodstuffs and other agricultural products and in handicrafts, and the movement of persons among the countries, have gone on without benefit of international agreement and often in disregard of law. Such traffic would certainly be facilitated if it were no longer subject to official harassment. On the other hand, harmonizing economic policies among the West African countries would tend actually to reduce their trade among themselves, since much of it is conducted to take advantage of differences in the prices officially stipulated for cash-crops, agricultural inputs and imported foodstuffs, or to avoid official controls on the convertibility of national currencies. In addition, a common external tariff would remove the West African trade presently actuated by differences in national customs duties.

Such considerations were not the basis of ECOWAS. As was said in the earlier discussion of the case for an African customs union, the principal gains anticipated lie in trade diversion for the benefit of new economic activities, especially in large-scale manufacturing, which are established under the protection of the common external tariff. For some considerable time, these gains might not be large. Transport and communications are relatively undeveloped in West Africa, and the costs of marketing goods and services are therefore high. The market effectively available to most enterprises would consequently be much smaller than is suggested by estimates of the combined populations or GDPs of the sixteen states. In the immediate future, there are probably few industries whose scale of operation and unit-costs would be much affected if they had access to a West African common market instead of only a national or other sub-regional market.

High internal transport costs could, of course, be offset by more protection. Trade diversion could be made greater by a higher common external tariff. But there is bound to come a point at which the price paid for trade diversion is questioned – by consumers in West Africa if not by theorists of economic growth.

Such trade diversion as was achieved would be politically contentious. The location of the new activities is crucial. If they are put where they can operate least inefficiently, some countries will be denied them, and must therefore forgo the growth-inducing effects claimed for them – as well as losing the opportunity to buy their products more cheaply from outside the union. These countries might be financially compensated for their loss. Or the Community might adopt a policy of distributing new activities 'equitably' among its members. In either case, the reduction in the costs of protection, which is the avowed rationale of the customs union, would be nullified at least in part. These difficulties are not insurmountable; if they were, customs unions (including national customs unions) would never have been created. But they do suggest that, for political as well as economic reasons, progress toward implementing the Treaty of Lagos and realizing the gains promised by it is likely to be slow. No early transformation of the composition and direction of the external trade of West African states is to be expected from the founding of their Economic Community in 1975.

Summary

West African export performance in recent years should be judged by reference to the composition of exports, and changes in their relative prices, as well as to trends in volumes. While export composition has changed little in some countries, in others it has been greatly altered – above all in Nigeria, because of the emergence of mineral oil as the predominant export. Whereas minerals made up less than one-fifth of the total value of West African exports in 1960, by 1979 their contribution had reached about three-quarters. Whereas Nigeria accounted for 36 per cent of total West African exports in 1960, by 1979 its share was over 74 per cent.

The changing relative prices of exports are summarized in estimates of the terms of trade. In the 1960s, the commodity terms appear to have improved for all West African countries except the iron ore exporters; in the 1970s, they deteriorated for many of the countries but improved greatly in Nigeria, Ghana, the Ivory Coast and Togo. The income terms improved for all the countries during the first period, in many cases briskly. In the second period, there was more variety of experience. The purchasing power of exports grew markedly in Nigeria, the Ivory Coast and some smaller countries, but contracted in four or five other countries. Measurements for West Africa as a whole are strongly influenced by the experience of Nigeria, the Ivory Coast and Ghana, which supplied about 70 per cent of the region's exports in 1960 and 90 per cent in 1979. For West Africa as a whole, the trend in the commodity terms was mildly favourable between 1960 and 1972 and highly favourable from 1973 to 1979, while the purchasing power of exports rose rapidly throughout the period; but the years immediately following 1979 were much less favourable.

The composition of West African imports has altered because of changes in the structures of domestic production and expenditure and changes in import costs, particularly the cost of imported fuel. The food import ratios appear high for countries that use so much of their labour in agriculture, but the statistics do not suggest an upward trend. The ratio of fuel imports rose dramatically for many countries after 1973 (and will have risen again after 1978), but in Nigeria this ratio has of course fallen, and it can be expected to fall in the Ivory Coast in the 1980s. For the region as a whole, the most striking change has been the increase in the relative importance of machinery and transport equipment – from 24 per cent of total imports in 1960 to 40 per cent in 1978. A comparison of Nigeria and Ghana shows that considerable changes are possible in the uses made of imports whether or not a country is experiencing rapid economic growth.

Nigerian oil has made the export ties of West Africa as much transatlantic as with Western Europe. The import ties are more largely with the European Community (about 56 per cent of 1979 imports). There has been considerable diversification of trading connections by the standards of earlier times. In every country, the trading predominance of the former metropolitan power has been reduced since 1960. Intra-African trade is still relatively insubstantial; only for a few of the Sahelian countries does it exceed 15 per cent of total exports or imports.

A survey of the external debt problems of West African governments suggests that they are fundamentally attributable to financial irresponsibility and administrative incoherence, with unforeseen negative movements in the terms of trade (as in Ghana and Sierra Leone in the early 1960s, and in the Ivory Coast, Liberia,

Mauritania, Sierra Leone and Togo at the beginning of the 1980s) acting only as precipitants. Of particular importance in creating insupportable debt burdens have been the enterprise of foreign salesmen offering supplies and contracting services on rather short-dated credit terms, the proliferation of parastatal bodies operating without centralized control, and the establishment of autonomous monetary systems allowing recourse to the deficit financing of governments. Membership of the *Union Monétaire Ouest Africaine* has provided protection against the last of these hazards to several countries, but has not been a sufficient obstacle to debt accumulation. Petroleum has saved Nigeria from serious problems of external debt (though the country became a substantial borrower in the Eurocurrency market in 1978), but the official reactions to volatility in the earnings from the mineral have produced great instability in imports, a steep upward trend being checked by administrative measures in 1979 and again in 1982.

The official values put on West African currencies diverge from market values in proportion to the stringency with which the government controls external transactions and the vigour with which it has inflated domestic purchasing power. At one extreme lie the almost valueless Guinean syli and Ghanaian cedi; at the other, the CFA franc used in the UMOA and the US dollar used in Liberia. The Nigerian naira has been the most important currency overvalued at the official rate. Correcting an overvaluation is politically difficult and, as Ghanaian experience has shown, futile unless accompanied by monetary discipline.

Foreign trade regimes were liberal in West Africa until the depression of the 1930s and the Second World War led to the creation of the franc zone and the sterling area. Political independence seemed to offer in West Africa the alternatives of escaping from that enforced bilateralism toward a more general incorporation into the world economy, or of resisting such transoceanic integration in favour of the development of African unity. The latter alternative implied trade diversion, but this was argued to be a long-run gain through its effects on the determinants of economic growth. In practice, the alternatives were found not to be mutually exclusive and both were pursued. The explanations are that West African exports were mostly non-competitive with the production of the region's principal trading partners in Western Europe, and that the governments of the latter did not insist on the principle of reciprocity in their trade agreements with Africa.

On the one hand, therefore, West African countries have obtained free or preferential access for their exports to the markets of the EEC – initially the francophone countries under the Treaty of Rome and the Yaoundé Conventions of 1962 and 1969, later all the countries of the region under the Lomé Conventions of 1975 and 1979. Pressure by the governments of countries thus associated with the EEC for the guaranteeing or stabilizing of their export earnings led eventually to the Stabex and Minex schemes. The effects in West Africa of these associations have been trade-creating at the expense of the former metropolitan powers. Their effects in Western Europe have perhaps been slightly trade-diverting at the expense of other developing countries. There has probably also been some aid diversion at the expense of other developing countries through the institution of the European Development Funds.

On the other hand, attempts have been made economically to integrate the independent states of West Africa by customs unions, the EEC members being prepared to forgo most-favoured-nation rights for this purpose. The most ambitious of these attempts, propelled by a government concerned to establish Nigerian claims to political leadership in Africa, resulted in the founding in 1975 of the Economic Community of West African States, which now embraces all sixteen countries in the region. The implementation of the ECOWAS treaty in the 1980s faces daunting political obstacles, and the principal gain envisaged from implementation – trade diversion toward large-scale industry operating on the basis of the regional common market – is unlikely to be immediately large. No early transformation can be expected in the composition and direction of West Africa's external trade examined in the earlier sections of this chapter.

5
Policy Instruments

Introduction

Earlier chapters have examined the anatomy of West African economies under the headings of economic structures, population and labour force, and external trade. Attention now turns to the physiology of these economies. It would be impossible to explain their workings without consideration of government interventions, controls and policies. Actually to make policies pivotal in the exposition has certain advantages – in bringing out the contrasts between private economic actions, which are for the most part economically motivated, and government economic actions, which frequently are otherwise actuated, and in allowing some conclusions to be drawn about how the declared objectives of government economic policies might best be fulfilled.

These declared objectives do not vary significantly among countries. In Nigeria, for example, they are the establishment of 'a united, strong and self-reliant nation, a great and dynamic economy, a land of bright and full opportunities for all citizens, and a free and democratic society'. Conditions for achieving these objectives have been specified to include increases in income per head, a less unequal distribution of income, reduction in unemployment, greater supplies of human skills, further economic diversification, wider dispersion of economic opportunities, and increased indigenous participation in the ownership and control of economic enterprises.[1] Neither these aspirations nor the means specified for achieving them would be denied elsewhere in West Africa; only the emphases might vary. The latent purposes of government are, of course, different. Public servants are mainly concerned to retain their jobs, to obtain promotion, to enlarge their authority. Rulers, ministers and legislators want primarily to stay in power (or occasionally to give it up without becoming dangerously exposed to the actions of their successors).

The results of executing government economic policies will be mainly considered in Chapter 6. The present chapter is focused on policy instruments, the institutional means available to governments for implementing their policies. Governments require means of securing resources for official use, of using for official purposes the resources so obtained, and of controlling or influencing use of the resources left in private hands. These interventions in resource allocation generally have inadvertent consequences or side-effects, the suppression or correction of which requires further instruments of policy; for example, retail price controls may follow import licensing, in an attempt to remove the opportunities for monopoly profit that have been created. West African governments have sometimes been extremely ambitious in their economic policy-making. Especially in the early years of independence, many politicians believed governments to have almost unlimited effective powers over economic life. The capacity of the policy instruments and institutions actually to deliver the intended results is therefore of great practical importance. So is the possibility that, because of administrative incapacity or for other reasons, the results intended may be transmuted in the course of executing policies.

Securing resources for official purposes

Governments secure resources for official uses from taxation, the earnings of government-owned property, borrowing, and external aid. The ratio to the GDP estimate of the recorded expenditure of the central or national government is a rough indication of the extent of this resource use (though not a complete indication, since resources may also be harnessed independently by extra-budgetary organizations within the central government, by subordinate – or, in the Nigerian federation, coordinate – units of government, and by parastatal bodies). In 1977, this ratio appears to have been between 15 and 20 per cent in Benin, Ghana, Mali, Niger and Upper Volta; between 20 and 25 per cent in Liberia, Senegal and Sierra Leone; and over 33 per cent in The Gambia, Guinea-Bissau, the Ivory Coast, Mauritania, Nigeria and Togo.[2] There is, of course, a correspondence between these measurements and the ratios to GDP of capital formation and government consumption, which were discussed in Chapter 2, since much investment (usually 50 to 80 per cent) is on central government account and there is considerable overlap between government consumption and

Table 22 Central government budgetary operations, *c.* 1977 (percentages)

	Benin 1975	*The Gambia 1977/78*	*Ghana 1977/78*	*Liberia 1978/79*	*Mali 1977*	*Mauritania 1977*	*Niger 1976/77*	*Nigeria 1977/78*	*Senegal 1978/79*	*Sierra Leone 1977/78*	*Togo 1977*	*Upper Volta 1977*
Income taxes	14.6	9.8	21.4	29.2	20.0	12.9	25.7	60.6	17.9	22.3	33.6	12.9
Individual	*5.4*	*3.7*	*12.3*	*14.4*	*5.1*	*7.9*	*4.9*	. .	*5.5*	*5.7*	*2.3*	*8.3*
Corporate	*6.0*	*5.4*	*9.1*	*14.4*	*13.8*	*2.5*	*17.3*	*60.6*	*7.9*	*16.6*	*29.5*	*3.7*
Unallocated	*3.2*	*0.7*	–	*0.4*	*1.1*	*2.5*	*3.5*	–	*4.5*	–	*1.8*	*1.0*
Domestic taxes on goods and services	15.0	2.1	29.3	19.2	18.7	10.8	29.3	2.2	26.3	16.3	14.9	18.1
General	*5.8*	–	*5.2*	*2.3*	*15.2*	*8.7*	*23.8*	–	*13.4*	–	*11.2*	*6.6*
Excises	*6.3*	*0.9*	*23.2*	*8.0*	*2.3*	*2.0*	*4.4*	*2.2*	*10.9*	*14.4*	*3.2*	*5.4*
Other	*2.9*	*1.2*	*0.9*	*8.9*	*1.1*	*0.1*	*1.1*	. .	*2.0*	*1.9*	*0.5*	*6.0*
Taxes on international trade	54.5	42.1	38.6	35.4	29.5	21.4	18.7	13.0	40.8	51.4	39.2	48.9
Import duties	*45.8*	*36.4*	*17.3*	*34.8*	*18.0*	*20.9*	*13.2*	*13.0*	*39.1*	*34.9*	*16.0*	*43.0*
Export duties	*5.2*	*5.6*	*21.3*	*0.5*	*11.1*	*0.5*	*4.6*	. .	*1.6*	*16.0*	*16.7*	*3.0*
Other	*3.6*	–	–	*0.2*	*0.3*	–	*0.9*	–	–	*0.5*	*6.5*	*2.9*
Other taxes	5.6	0.7	0.5	2.9	16.8	4.3	10.0	0.1	4.7	2.1	7.0	3.0
Non-tax revenues	10.3	26.1	10.1	3.0	7.2	17.5	14.9	24.1	10.0	7.9	5.4	14.3
Grants	–	19.2	0.1	10.3	7.8	33.2	1.4	–	0.3	–	–	2.7
Total, revenues and grants	100.0	100.0	100.0	100.0	100.0	100.0	100.0	100.0	100.0	100.0	100.0	100.0

Overall surplus/deficit as % of total expenditure (incl. net lending)	18.1	−25.9	−57.7	−38.6	−0.5	−22.8	12.4	−24.2	1.0	−10.2	−15.0	5.5
Financing:												
external	n/a	19.8	2.0	35.2	−1.6	8.3	0.1	3.4	n/a	n/a	0.7	n/a
domestic, non-bank	n/a	7.3	8.0	0.9	0.2	3.7	−0.4	n/a	n/a	n/a	−0.2	n/a
domestic, bank	n/a	−1.2	47.7	2.6	1.9	10.7	−12.2	n/a	n/a	n/a	14.5	n/a

Notes: n/a indicates not available.

. . indicates less than 0.05%.

Figures for The Gambia, Liberia, Mali, Mauritania, Niger and Togo rest on data for consolidated central government, i.e. including accounts of extra-budgetary organizations within central government; figures for Benin, Ghana, Nigeria, Senegal, Sierra Leone and Upper Volta rest on data for budgetary central government only.

Source: calculated from data in *Government Finance Statistics Yearbook*, vol. 4, 1980 (Washington DC: International Monetary Fund).

government current expenditure. Of most significance are the high government expenditure ratios of Nigeria (40 per cent) and the Ivory Coast (37 per cent); ironically, both countries are generally regarded as having provided institutional environments particularly favourable to private enterprise. The highest ratio, 55 per cent in Mauritania, reflects that country's involvement in 1977 in the war in the Western Sahara, as well as large government investment programmes.

Table 22 shows, for twelve countries of the region in or about 1977, the relative importance of the principal categories of taxation, non-tax revenues, grants received (almost entirely from abroad), and borrowings to cover a budget deficit (or repayments to dispose of a surplus).

A characteristic of these revenue structures is the preponderance of receipts from levies on international trade, and sometimes from the taxation of one or a very few industries producing largely or entirely for export. To take first the most important case, the federal government of Nigeria in 1977/78 raised 60.6 per cent of its revenue (including revenue subsequently transferred to the governments of the States) from taxation of corporate income, 13.0 per cent from import duties, 2.2 per cent from excises, and relatively negligible amounts from other taxes. The remaining 24.1 per cent of federal revenue came from non-tax sources. The budget deficit was equal to 24.2 per cent of total federal expenditure (including lending to State governments and other bodies), and it was largely financed from domestic sources.

This remarkable revenue structure resulted from the appropriation by the government of most of the enormous economic rent generated in oil extraction. In fact, the greater part of the corporate income taxes (54 per cent of total federal revenue) consisted in the proceeds of the petroleum profits tax levied on the oil mining concessions, and the greater part of the so-called non-tax revenue (17.2 per cent of total federal revenue) was contributed by mining rents and royalties, which might reasonably be regarded as another form of taxation of mining (and are so regarded in the Nigerian official statistics). Thus, about 71 per cent of total federal revenue in 1977/78 was derived directly from the oil industry – and over 76 per cent of total federal tax revenue, if mining rents and royalties are regarded as taxes.

It must be added that personal income taxes appear so negligible in the federal accounts because they are levied almost entirely by the State governments. Even so, these and other independent State revenues have been of minor importance in Nigeria since the mid-1970s; in 1977/78 they may have equalled about 5 per cent of the federal revenues.

Revenues derived directly from oil became preponderant in Nigeria during the 1970s; from 1974 they averaged over 75 per cent of annual federal revenue. Previously, import and export duties had been the principal revenue sources; in 1960, for instance, they contributed together nearly three-fifths of the combined revenues of the governments of the federation.

Mining is of less direct importance as a source of revenue in other mineral-exporting countries because the margins between sales value and cost of production are much less for other minerals than they have been for oil. The influence of a relatively large mining sector can nevertheless be seen in the share of total revenue contributed by corporate income taxes in Liberia, Niger, Sierra Leone and Togo. In Niger, there are also substantial royalties included in non-tax revenue; uranium mining was the source of some two-fifths of total revenue by the beginning of the 1980s. In Sierra Leone, mining has contributed to the proceeds from export duties (through an export tax on the sales of the small-scale diamond producers) as well as to corporate income taxation. The public share in the proceeds of Mauritanian mining has been obtained mainly through non-tax revenues, turnover taxation and import duties; in some years it may have provided as much as one-half of domestic revenue. For Guinea, which does not appear in Table 22, the dependence of the revenue on bauxite and alumina production could well be as great.

Except for Niger, Nigeria and Togo, the table shows income taxes (individual and corporate together) contributing about one-fifth at most of total revenue. This low proportion attests to the limited scope of incorporated economic enterprises in many of the countries, and to the practical difficulties experienced in them all of taxing personal incomes, other than the salaries of the comparatively small numbers of people in formalized and relatively well-paid employments, where the tax can be withheld at source.

Taxes on international trade (either import duties alone, or import and export duties together) are the principal domestic source of revenue everywhere except in Niger and Nigeria – and even those exceptions result from taxation of domestic production which is intended entirely or largely for export. The importance of import and export duties derives less from the substantial foreign trade ratios of West African countries than from the facility with which these taxes could be collected, where external trade was largely seaborne and concentrated in small numbers of trading organizations.

General sales, value-added and turnover taxes are used in the francophone countries, Ghana and Liberia, but many of the taxed products are simply imports or exports, so that, to some extent, these levies are only formally distinguishable from taxes on international trade. Excises have become important in a few countries, notably Ghana, reflecting the shift in the end-use of imports toward materials and semi-products for use in local industries; they are an attempt to recoup revenue lost through reduction in the importation of finished goods. In Liberia, the category of other domestic taxes on goods and services is relatively large (8.9 per cent) because it includes the vessel registration fees charged for the so-called flag of convenience; and in Upper Volta (6.0 per cent), because it includes the profits of a fiscal monopoly in tobacco.

The residual tax category in the table includes licence fees (for example, motor vehicle licences); poll taxes in the countries (Benin, Liberia, Mali, Niger) where they are collected centrally; social security contributions and payroll and property taxes, chiefly in the francophone countries; and unallocated revenues which are sometimes substantial. Non-tax revenues include net receipts of public enterprises and funds (most notably, 20 per cent of Gambian revenues in 1977/78 came from the Produce Marketing Board), central bank profits, mining royalties, earnings on government investments and loans, fines, and charges for services and non-industrial sales (such as medical services).

The most important omission from the table is the Ivory Coast. In that country, income taxes were reported to have contributed about 21 per cent of budgetary revenue in 1976, import and export duties 55 per cent (38 and 17 per cent respectively), and value-added and excise taxes 19 per cent.[3] Again, therefore, the revenue depended heavily on the taxation of external trade.

The taxation of exporting

Reliance on import taxation was a feature of colonial administration. Not only were import duties easily and therefore cheaply collected. In addition, the imports were largely of novel consumer goods for which the income-elasticities of demand were high and the price-elasticities low. Imports were therefore peculiarly suitable as objects of taxation in the early colonial period. On the other hand, the taxation of exports was to be sparing, if not altogether avoided. Export taxes could not usually be passed on to foreign buyers. They would depress the profitability of large exporting enterprises, like the mining companies, and they would deter the movement of African farmers into cash-cropping for export. Importing capacity would thus be reduced, and with it the flow of revenue from import duties.

This view of export taxation changed as a result of the establishment in British West Africa, early in the Second World War, of statutory monopolies of export crop sales. The manifest reasons for this innovation were prevention of a collapse in the local price of cocoa, consequent on the loss of the German market, and encouragement of production of vegetable oils and oilseeds after the loss of South-East Asian supplies in 1942. In neither case was monopolization of supplies a necessary, or indeed an appropriate, instrument. Farmers were, in fact, paid much less than the prices (net of marketing costs) obtained for their crops in Great Britain, and the monopolies accordingly realized surpluses on their operations, which were then lent to the British government. In addition to this enforced lending by African farmers, a latent purpose of the market control was to avoid competition with and among the expatriate merchant firms established in West Africa; they were the original buying agents licensed by the statutory monopolies, and they were assigned percentage buying quotas and guaranteed margins of profit on their transactions.[4]

After the war, the surpluses accumulated by the export monopolies could not, in practice, be apportioned among the farmers who had earned them, yet had somehow to be bestowed – and without prejudice to the British balance of payments. Dismantling the wartime marketing arrangements was held to mean abandoning 'going concerns' and a wanton disregard of the 'success' they had achieved. The solution was to replace the temporary wartime institutions by peacetime marketing boards having the same monopolistic powers over the purchase, local pricing and sales to users of export crops. Such boards were established in the late 1940s for cocoa, palm produce, groundnuts and cotton in Nigeria, for cocoa in the Gold Coast, for a range of export crops in Sierra

Leone, and for groundnuts in The Gambia. They inherited the wartime surpluses and had the power of adding to them. Their avowed purpose was stabilization.

Economic stability was prized in the 1940s because it had been so conspicuously lacking before the war. In the guise of a full-employment equilibrium, it had emerged as the supreme aim of economic policy-making in Great Britain and elsewhere; security in employment was primary among the welfare that electorates demanded and that governments of the time undertook to deliver. Colonial peoples were deemed to be also deserving of welfare and, if stability could not be provided for them mainly in employment because relatively few of them worked for hire, a counterpart could be found in a sustained level of aggregate money income, or of major components of this income, such as might be attained by the obligatory collective marketing of export crops.

As a monopoly, a marketing board could stabilize the producer price of a crop merely by stipulating a uniform payment per ton or head-load to the agents it licensed. Stabilizing the producers' aggregate money income from the crop was not so simple; it required that this uniform price should be adjusted between seasons in the same proportion as the marketed quantity changed, and in the opposite direction to that change. While such adjustments would, in principle, provide equal seasonal incomes to the growers in the aggregate, the marketing board, reselling the produce abroad in a free market, would experience fluctuations in its earnings. In consequence, the board might either obtain a surplus or suffer a deficit in its trading operations.

Thus, in Figure 3, if OA, OB and OC represented the quantities purchased by the board in three successive seasons, the setting of the producer price at OF, OG and OK respectively would provide equal seasonal incomes to the growers in the aggregate (the points M, H and P lie on a rectangular hyperbola). If the relationship between demand in the export market and prices net of marketing costs were such as is represented by the line DE, the quantity OA in the first season would sell at OJ (higher than the producer price OF) and the quantity OC in the third season would sell at ON (lower than the producer price OK). In the first season there would therefore be a marketing board surplus measured by the area FJLM, and in the third season a deficit measured by KNRP.

Figure 3 Intended *modus operandi* of a marketing board

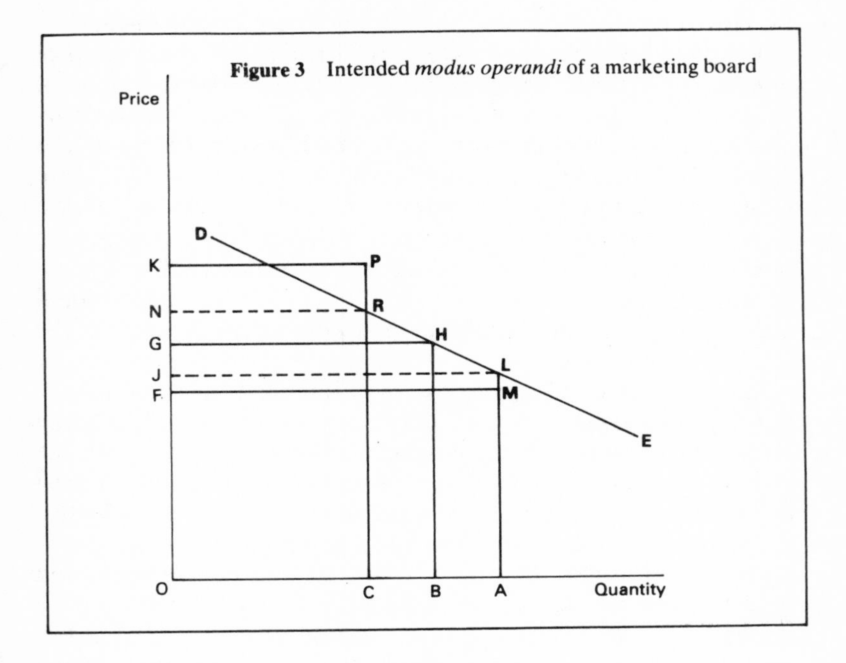

The intended *modus operandi* of the marketing boards was that the surpluses of some seasons, kept as stabilization reserves, should finance the deficits of other seasons, and that the average producer price over a period of years should approximately equal the average net price realized in the export market. In the longer term, therefore, the producers in the aggregate would receive substantially what they had earned, but in the shorter term the level of their income would be protected by breaking the direct link between world market prices and local producer prices.

It may well be thought astonishing that a scheme of this nature should not only have been sanctioned by the British government but also have found support among economists.[5] Even if it had functioned according to design, it would merely have done for producers in the aggregate what they might individually have chosen to do for themselves, but were now prevented from doing[6] –namely, to defer to bad years the extra earnings of good years. In practice, the scheme could not function according to design. Since intraseasonal stability in the producer price was considered no less desirable than (and tended to be confused with) interseasonal stability in producer income, and since payment by instalments was usually impracticable, the producer price had to be fixed in advance of each harvest and therefore in ignorance of both the volume of marketed output and the average price it would realize when sold by the marketing board (in terms of Figure 3, neither the quantities OA, OB, OC nor the position of the demand curve DE were known). Consequently, the power of a board to fix a uniform producer price could have no certain implication for aggregate producer income, either in one season or when averaged over years; far from being a condition of stability in that income, it could actually produce more instability than would have occurred in the absence of the monopoly.

Faced by these obstructions to their intended *modus operandi*, the postwar marketing boards, like their wartime precursors, erred on the side of caution, fixing producer prices usually well below the average market prices that were subsequently realized. Several forces sustained this tendency – the continuing need of the British government to borrow from the colonies, the belief that a postwar boom in the prices of the crops controlled by the boards was bound soon to collapse, the inclination to judge the success of statutory marketing by the financial wealth that was amassed. So considerable became the surpluses between the realized proceeds of the export crops and the earnings paid to producers that, within a few years of the establishment of the peacetime boards, colonial governments (now passing under the control of elected politicians) perceived that these surpluses could as well accrue as tax revenue as be allowed further to swell stabilization reserves on which few drawings were made. Overt taxation in the form of export duties (and, in Nigeria, of produce sales taxes imposed by the Regional governments) therefore partly replaced the covert taxation consisting in the trading surpluses of the marketing boards. Later, the accumulated surpluses themselves were substantially made over as grants or loans to the government of Ghana and the Regional governments in Nigeria. By 1954, the marketing boards in those countries had become fiscal instruments of government rather than independent statutory authorities, and their purpose the deflection of purchasing power from those who had earned it to those who were supposed to know better how to use it – principally the government, but also the boards themselves as costly systems of administration and the licensed buying agents who continued to earn guaranteed margins on their transactions.

Stabilizing institutions did not become established in French West Africa until the mid-1950s. These territorial *caisses de stabilisation* stipulated producer prices to be paid for the export crops, and were intended to reduce the interseasonal fluctuations in those prices by alternatively charging levies or paying subsidies on exports. They left room for private exporters within a framework of guaranteed cost allowances and profit margins, but in some countries (Guinea, Mali, Niger, Senegal, Togo) exporting was taken over by state agencies soon after independence. Being established nearly a decade after the marketing boards, the *caisses* did not immediately experience, as the boards had done, a phase of rapidly rising export prices. Their initial funding had therefore to be partly provided by France. In addition, the French *surprix* arrangements for coffee, palm produce, bananas and groundnuts enhanced the profitability of exporting those crops, and thus helped to support the producer prices. This external support continued until 1969 through the aid earmarked for agricultural efficiency under the first Yaoundé Convention. Until that date, the gap between producer and export prices was usually large, as in the formerly British territories, but not so much because the producers were being squeezed as because the export prices were held up by French and EEC subsidies, allowing revenue from the exports to accrue both to the stabilization funds and directly to the African governments. After 1969, receipts from exports continued to be appropriated by public authorities, but now at the expense of African farmers rather than European consumers. As in the formerly British territories, the funds supposedly amassed in order to stabilize farmers' aggregate incomes were more largely used in financing government expenditures.

The marketing boards and stabilization funds have certainly removed intraseasonal fluctuations in producer prices, and they have on occasion protected those prices from the more violent downswings in world market prices. But stabilization in the sense of balancing underpayments to farmers in one period by overpayments in another has never been seriously attempted by any of these institutions. It was entirely in keeping with prevailing norms that, when an export marketing monopoly was introduced in Liberia in 1962, no intention of eventually returning its trading surpluses to the producers was even professed; they were simply to be divided between the government of Liberia and the foreign corporation which managed the monopoly.[7]

Table 23 Proportions of export-crop producers' earnings retained

A. AGGREGATE PRODUCER RECEIPTS OVER PERIOD AS PERCENTAGE OF NET EARNINGS				
Ghana	Cocoa	1947/48–1964/65	56	(a)
Nigeria	Cocoa	1947/48–1961/62	68	(b)
	Cotton	1949/50–1960/61	78	(b)
	Groundnuts	1947/48–1960/61	75	(b)
	Palm kernels	1947–61	72	(b)
	Palm oil	1947–61	79	(b)
Senegal	Groundnuts	1962/63–1972/73	61	(c)
B. PRODUCER PRICE AS PERCENTAGE OF NET FOB PRICE (ANNUAL AVERAGE)				
Ghana	Cocoa	1971/72–1979/80	44	(d)
Ivory Coast	Cocoa	1965/66–1974/75	50*	(e)
	Coffee	1965/66–1974/75	54*	(e)
Nigeria	Cocoa	1964/65–1976/77	55*	(f)
	Groundnuts	1964/65–1976/77	52*	(f)
	Palm kernels	1964/65–1976/77	63*	(f)
	Palm oil	1964/65–1971/72	61*	(f)
Mali	Cotton	1971/72–1979/80	48 (6)	(d)
	Groundnuts	1971/72–1979/80	48 (6)	(d)
Senegal	Groundnuts	1971/72–1979/80	57 (8)	(d)
Sierra Leone	Cocoa	1968/69–1973/74	53	(g)
	Coffee	1968/69–1972/73	60	(g)
	Palm kernels	1949–63	77	(h)
		1968/69–1973/74	70	(g)
Togo	Cocoa	1971/72–1979/80	39	(d)
	Coffee	1971/72–1979/80	34	(d)
	Cotton	1971/72–1979/80	70	(d)

Notes: * Percentage of gross FOB price.
(6) Average of 6 observations (years) only.
(8) Average of 8 observations (years) only.

Sources: (a) Björn Beckman, *Organising the Farmers: Cocoa Politics and National Development in Ghana* (Uppsala: Scandinavian Institute of African Studies, 1976), Appx A, Tables 1 and 4.
(b) G. K. Helleiner, *Peasant Agriculture, Government and Economic Growth in Nigeria* (Homewood, Ill.: Richard D. Irwin, 1966), Appx, Tables V-F-1 to V-F-5.
(c) *Senegal: Tradition, Diversification and Economic Development* (Washington DC: World Bank, 1974), Table 2, p. 12.
(d) *Accelerated Development in Sub-Saharan Africa* (Washington DC: World Bank, 1981), p. 56.
(e) *Ivory Coast: the Challenge of Success* (Baltimore & London: Johns Hopkins University Press for the World Bank, 1978), Table 4.9, p. 81.
(f) Robert H. Bates, *Markets and States in Tropical Africa* (Berkeley & Los Angeles: University of California Press, 1981), Appx B.
(g) John Levi, *African Agriculture: Economic Action and Reaction in Sierra Leone* (Farnham Royal: Commonwealth Agricultural Bureaux, 1976), Table 4.13, p. 192.
(h) Ralph Gerald Saylor, *The Economic System of Sierra Leone* (Durham, NC: Duke University Press, 1967), Table 17, p. 114.

The extent to which export-crop producers have been taxed, both overtly and covertly, is indicated by the ratio of their aggregate receipts over a period of years to the aggregate FOB sales value (net of purchasing costs) of their crops, as in Table 23A. Alternatively, the ratio of the average producer price to the average realized export price can be taken – again on the basis of a period of years, so as to allow for the possibility of the marketing institution actually performing some interseasonal stabilization. Ideally, the realized export price should also be measured net of the costs of the marketing institution. In the instances in Table 23B where this deduction from the FOB price has not been made, the percentages shown understate by several points the proportion of the unit-value retained by the producers.

It appears from the evidence assembled in the table that the levies imposed on the incomes of export-crop farmers have generally been heavy. In no instance shown in the table do they amount to less than one-fifth of the sums seemingly available for remuneration of the farmers; more usually, they amount to about two-fifths, and there are two instances where they exceed three-fifths. It should be observed that these are average, not marginal, tax rates and that they apply to all producers of the crops concerned, regardless of their personal circumstances.

The true extent of the withholdings from producers' incomes has often been greater than these measurements suggest. One reason is that avoiding competition in crop purchasing can be expected to have inflated purchasing costs at the expense of producer prices. Where competition has been altogether absent, as in Ghana in the early 1960s,[8] farmers have not even necessarily received the price supposedly guaranteed them. The practice of paying farmers by 'chits' or promissory notes, the cash-value of which must be problematic, has also been reported at times in Ghana, Senegal and Sierra Leone. In addition, the overhead costs of the marketing boards and stabilization funds are generally believed to be extravagant, since these institutions serve political as well as commercial purposes – in some countries, they have been popularly regarded as first among the predators on the farming communities.

Further, the earnings in local currency from export crops, and therefore the amounts available for payment to farmers, would often have been greater at equilibrium exchange rates than at those officially maintained. In other words, those earnings have been taxed implicitly by currency overvaluations, as well as overtly by export duties and covertly by the surpluses of the marketing institutions. As was mentioned in Chapter 4, overvaluation of the currency has been pronounced in Ghana (especially since 1974), in Nigeria since the civil war, and in Sierra Leone since 1975; in the countries using the CFA franc, it has been much less, because of the free convertibility of that currency with the French franc, though not entirely absent. In Ghana in the late 1970s, farmers were retaining about 40 per cent of the net FOB price of cocoa converted to cedis at the official rate. Assuming overvaluation of the cedi by a factor of two (which is probably a very conservative assumption), the proportion retained by farmers of the export price converted into cedis at an equilibrium rate of exchange would be only 20 per cent. In other words, four-fifths (at least) of their earnings were deducted at source.

It is true that the proceeds of an implicit exchange tax arising from currency overvaluation are distributed as implicit subsidies on imports and foreign factors of production. The implicit tax on export-crop farmers may be offset in this way, but it is unlikely that the offsetting is anywhere near complete. Where overvaluation is pronounced and is accompanied by import restrictions, the principal beneficiaries of the subsidies are traders and manufacturers receiving import licences, institutions and persons obtaining permission to make transfers and factor payments abroad, and possibly the better-off urban consumers having access to stores in which price controls are effective. In those circumstances, not much of the advantage of cheap foreign exchange percolates as far as farming households.

The consequences of imposing taxes on export-crop farming which are both heavy, or extremely heavy, and usually discriminatory (since many alternative activities of the farmers are untaxed) will be considered in Chapter 6. For present purposes, it is enough to observe that the relative importance of export taxation in West Africa is greater than might be inferred from Table 22. The table isolates the contribution of export duties, but the surpluses of statutory export marketing institutions are concealed in non-tax revenue, or left altogether out of account where the operations of these bodies have not been consolidated with the budget of the central government. The facility with which taxes on exports can be collected (as compared with, for instance, a tax on personal incomes, a land tax, or a general sales tax) is an important explanation of the importance they have assumed. Ironically, this facility depends on the creation in the late colonial period of institutions supposedly designed to protect the interest of the export producers.

Foreign aid

Table 22 also understates the importance of external grants as a means of securing resources for West African governments; it shows these grants to have been an appreciable source of revenue only in The Gambia, Liberia, Mali and (especially) Mauritania. In Table 24, the grants included in Table 22 are listed as dollar magnitudes and compared with the grants reported by donors to have been made to the twelve countries concerned in the same, or approximately the same, years. The discrepancies are obviously very large. For the twelve countries together, the total reported by the donors is more than six times as great as that derived from the

Table 24 Comparison of reported grants to West African countries, *c.* 1977

	IMF Government Finance Statistics Yearbook		*OECD*	
	Year	*$ million*	*Year*	*$ million*
Benin	1975	–	1976	32.9
The Gambia	1977/78	9.0	av. 1977–8	16.3
Ghana	1977/78	0.8	av. 1977–8	59.9
Liberia	1978/79	23.0	av. 1978–9	31.1
Mali	1977	10.6	1977	73.8
Mauritania	1977	62.8	1977	130.6
Niger	1977	2.0	1977	65.5
Nigeria	1977/78	–	1977	37.7
Senegal	1978/79	1.5	av. 1978–9	164.6
Sierra Leone	1977/78	–	av. 1977–8	18.2
Togo	1977	–	1977	36.6
Upper Volta	1977	3.3	1977	79.4
Total		113.0		746.6

Note: Figures from the *Government Finance Statistics Yearbook* have been converted from local currencies to US dollars at contemporaneous official rates of exchange.

Sources: *Government Finance Statistics Yearbook*, vol. 4, 1980 (Washington DC: International Monetary Fund), and *Geographical Distribution of Financial Flows to Developing Countries* 1976/1979 and 1977/1980 (Paris: OECD, 1980 and 1981)

Table 25 Disbursements of official development assistance to West African countries, 1978–80 (annual averages)

	Benin	*The Gambia*	*Ghana*	*Guinea*	*Ivory Coast*	*Liberia*
$ million						
ODA grants	59.2	26.9	66.3	34.1	102.6	36.3
ODA loans (net)	19.5	15.1	91.9	34.0	65.2	39.3
Total ODA (net)	78.7	42.0	158.2	68.1	167.8	75.6
Percentages						
ODA by DAC members	48	35	53	28	75	50
by multilateral agencies	50	50	34	67	25	40
by OPEC members	2	15	13	5	–	10
	100	100	100	100	100	100
by EEC & members	54	43	54	31	74	28
Ratio of ODA to total net financial inflows	42	78	78	62	24	15

budgetary statistics of the recipient governments. All twelve appear from the donors' reports as beneficiaries of grants, some of them (Senegal, Upper Volta) as very large beneficiaries; only for Nigeria could the grants be said to be relatively unimportant.

Although incompleteness in the government finance statistics may partly account for these discrepancies, the chief explanation is that many grants are provided in forms that do not show as revenue – as technical assistance, food, and cancellation of debts (in fact, about 45 per cent of the total of $746.6 million shown in Table 24 consisted in technical cooperation grants).

Table 25 presents a wider view of the magnitude of the external aid received by West African governments. It shows for each of fourteen countries the average annual dollar value of official

Mali	Mauritania	Niger	Nigeria	Senegal	Sierra Leone	Togo	Upper Volta
161.3	123.4	142.2	46.0	176.3	32.2	43.7	162.6
41.1	61.6	24.6	−10.9	89.4	28.6	57.3	27.5
202.4	185.0	166.8	35.1	265.7	60.8	101.0	190.1
52	23	60	49	57	54	62	67
43	33	36	51	41	43	38	33
5	44	4	–	2	3	–	–
100	100	100	100	100	100	100	100
51	34	58	35	64	41	67	57
95	105	68	4	73	90	85	95

Note: for definitions, see text.

Source: Geographical Distribution of Financial Flows to Developing Countries 1977/1980 (Paris: OECD, 1981).

development assistance (i.e. concessional transfers from governments and multilateral institutions, made for the promotion of economic development and welfare in the recipient countries) in the period 1978–80, divided between grants and loans net of amortization payments. The table also shows the proportions of the total ODA contributed by member-governments of the Development Assistance Committee of the OECD (almost all the 'Western' industrialized countries); multilateral agencies (including the World Bank and its affiliates, the European Development Fund, the African Development Bank and Fund, several Arab institutions, and the Technical Assistance agencies of the UN); the member-governments of OPEC; and the member-governments and collective institutions of the EEC (this last ratio being made up partly of bilateral DAC aid and partly of multilateral aid). (Aid to West Africa by the Russian and East European governments appears to have been relatively insignificant during the period under examination, and data on Chinese aid are not available.) Finally, the table shows ODA as a proportion of total net financial inflows (including non-concessional official transfers and private direct investment, bank lending and export credits as well as ODA) on average for the years 1978–80.

For the fourteen countries together, aid was received in this period at an annual rate of $1.8 billion, two-thirds as grants and one-third as loans (though these proportions were by no means uniform among countries). The grant element[9] in the loans was usually 50 per cent or more, but averaged much less (34 per cent) in the Ivory Coast. On the basis of the population figures given in Table 1, aid per head averaged $12.5 a year, but this figure rises to nearly $30 if Nigeria is excluded from the calculation. Perhaps unsurprisingly, aid to Nigeria was relatively negligible. The next largest countries, Ghana and the Ivory Coast, also fared relatively badly by this criterion, along with Guinea and Sierra Leone. At the other extreme lay the countries with the smallest populations – Mauritania (over $100 per head), The Gambia (over $70), Liberia and Togo (over $40) – along with Senegal. The mild negative correlation between aid per head and the population-sizes of countries is consistent with the view that aid is delivered to meet the needs of national governments, not peoples.

As a proportion of total net financial inflows, ODA averaged 46 per cent for the region as a whole in 1978–80. This proportion was relatively low for the Ivory Coast, Liberia and (especially) Nigeria. For half the countries, it exceeded 75 per cent. Just over one-half of the total ODA to West Africa was contributed bilaterally by the member-countries of the Development Assistance Committee (and about the same proportion was provided individually and collectively by the members of the EEC); nearly two-fifths came from multilateral institutions, and about one-twelfth from the member-countries of OPEC. The most striking departure from this pattern was Mauritania, the principal recipient of bilateral aid from Arab governments.

In Table 26 the net ODA disbursed in 1979 is shown in proportion to the GNP estimates for the fourteen countries, to their gross domestic investment and to their imports. According to these figures, aid constituted some 2 per cent of the combined

Table 26 Disbursement of official development assistance in relation to estimates of GNP, gross domestic investment and imports of West African countries, 1979

	Net ODA as percentage of		
	GNP	*Gross domestic investment*	*Imports*
Benin	10	47	26
The Gambia	24	125	26
Ghana	4	33	21
Guinea	4	24	29
Ivory Coast	2	6	7
Liberia	9	32	36
Mali	21	106	77
Mauritania	35	75	75
Niger	13	36	60
Nigeria	. .	. .	. .
Senegal	13	59	32
Sierra Leone	6	45	20
Togo	13	28	32
Upper Volta	20	96	99
14 countries together	2	6	10
14 less Nigeria	7	25	25
14 less Nigeria and Ivory Coast	9	41	35

Note: . . indicates less than one-half of 1 per cent.

Sources: for ODA, see Table 25; for GNP estimates, Table 1; for GDP, Table 6; for imports, Table 17.

West African GNP, and was equivalent to 6 per cent of the combined investment and 10 per cent of the imports of the region. These proportions do not appear remarkably large, and they would of course be less – about 1.5, 4.5 and 8 per cent respectively – if the loan-aid was measured as its grant equivalent.[10] But Nigeria, to which aid was negligible, and the Ivory Coast, to which it was of much less than average importance, together accounted for four-fifths of the combined GNP estimate, for nearly nine-tenths of the combined investment, and for nearly three-quarters of the import total in 1979. Omitting those two countries, ODA represented 9 per cent of the GNP of the remaining twelve, and was equivalent to 41 per cent of their combined investment and 35 per cent of their imports. Among individual countries, the ratio of ODA to GNP reached two digits in eight (all but one former French dependencies and long-term associates of the EEC), and the ratios of ODA to investment and to imports were particularly high in several Sahelian countries.

These measurements are far from exact and somewhat misleading. The difficulties in enumerating and the uncertainties of evaluating the components of GNP estimates were discussed in Chapter 2. The aid totals are records of outlays by the donors. Especially where the aid is given in kind or is otherwise tied to procurement from the donor,[11] its value in relation to the GNP is likely to be inflated, while the GNP itself is undervalued by its conversion to dollars at the ruling rate of exchange. The value of aid tends also to be exaggerated in relation to investment and to imports.[12]

Although subject to this caveat, the figures given in Table 26 (which may be read in conjunction with the negative resource balances estimated for 1979 and shown in Table 6) are such as strongly to indicate a heavy dependence on external aid of economic activity and external payments in most West African countries. Since the aid is donated almost wholly to governments, the dependence is particularly of public activity and payments; as some critics of aid have long observed, one of its effects is to extend the sway of government in the aided economies.[13]

It must be emphasized that what was true of most West African countries in 1979 was not true of the region considered as a whole. The reason is the relative economic size of Nigeria and the Ivory Coast, countries whose governments did not depend heavily on external aid. Elsewhere in West Africa, strong aid relationships appear to have developed as a result of attempts to discharge what are considered as appropriate responsibilities of national government in countries which are economically too small to sustain this task unaided.

For a national government and its agencies to be substantially financed by foreign governments and international institutions is perhaps the clearest manifestation of the dependency that is frequently said to characterize the economic functioning and public affairs of developing countries. In Nigeria, escape from this aid relationship was declared soon after 1960 to be a concomitant of political independence[14] – and it was achieved sooner than expected because of the rapid growth in the value of oil exports. Elsewhere, there have been no great expectations that aid could be dispensed with, but hopes have been entertained that it could be made less unreliable and unpredictable, less a matter of donors' choice and more a matter of recipients' rights.

The Stabex scheme described in Chapter 4 provided such rights to aid. The EEC members undertook to compensate, by grants or interest-free loans, for shortfalls in the earnings of ACP states from exports to the EEC. These rights were not unrestricted. They were qualified by the rules of the scheme and its commodity coverage, and also by the disregard of movements in the import prices of the ACP states and the limited financial provision made for Stabex in the European Development Fund.

Aid thus secured in conjunction with trade is nevertheless strongly attractive to a recipient government because of the automatic nature of its disbursement and the almost complete discretion that is enjoyed in its use. Aid of this sort, in contrast to most other sorts, is not dependent on the cooperation in and approval of some specific project by a donor agency or consortium of agencies. It is aid delivered as international purchasing power for general purposes rather than as prescribed goods and services for particular purposes. It can be used as the recipient, rather than the donor, thinks best.

Stabex is one way of obtaining aid through trade. The Sugar Protocol of the Lomé Convention is another: prescribed quantities of ACP sugar can be sold in the EEC market at prices which are higher than free market prices because of the heavy protection given EEC sugar producers. The French *surprix* arrangements mentioned in Chapter 4 are another illustration: in the 1950s and 1960s, prescribed quantities of certain African exports were purchased for the French market at prices in excess of those current in the world market. It is understandable that aided governments should often prefer to receive aid in such ways – as what is due to them, rather than as what donors see fit to give them.

International commodity agreements appear to offer another means of obtaining aid through trade. That was not the intention when they were legitimized by GATT. Their declared purpose was to limit short-run fluctuations in the prices of particular commodities. Minimum and maximum prices were therefore to be set (and revised from time to time) by agreement between exporting and importing countries, and the agreed price range would be defended by purchases for, or sales from, a buffer stock, by quantitative restrictions on exports or on production, or in both these ways. But, in principle, contact would not be lost between the agreed price range and long-run movements in the market equilibrium price resulting from changes in costs of production or shifts in demand.

A different purpose was attributed to international commodity agreements by the Final Act of the first United Nations Conference on Trade and Development in 1964; it was that of 'stimulating a dynamic and steady growth and ensuring reasonable predictability in the real export earnings of the developing countries so as to provide them with expanding resources for their economic and social development'.[15] International agreement was to be sought not for stable prices but for higher prices. The slogan 'trade, not aid', current at the time, meant aid obtained through trade.

That view of the objective of commodity agreements has since continued to be favoured by the governments of developing countries, including those of West Africa. It presupposes inelasticity in world demand for the controlled commodities (otherwise aggregate receipts would be reduced by higher prices). It also assumes solidarity among exporting countries, since demand is likely *not* to be inelastic for any one country acting independently (unless it supplies a large proportion of the world demand) and it therefore has a strong incentive to remain outside or to breach the agreement.

The effectiveness of an agreement in raising prices, and also its ability to survive, would be strengthened if not only sales but also purchases were controlled – as by importing countries making long-term commitments to buy not less than stipulated quantities. Such purchasing commitments figured in the 'integrated programme for commodities' canvassed by the UNCTAD Secretariat from 1974 as the leading feature of the proposed New International Economic Order.[16]

As has often been observed, a consequence of regulating production, and perhaps also consumption, in order to fix prices at acceptable levels is that innovation is deterred and economic structures become frozen. Such a process of immobilization is generally favoured by established producers, and not only in developing countries; so far as it is carried through, the costs are borne by prospective suppliers as well as by consumers.

West African governments have had interests in the international commodity agreements that have existed for tin (since 1953), coffee (since 1962), cocoa (since 1973), and rubber (since 1980). Only

Nigeria is concerned in the tin agreements; they have had little effect on prices and producers' earnings, and have perhaps owed their durability to their unimportance.[17] Purchasing for an international buffer stock of natural rubber began at the end of 1981, but had little immediate effect in raising prices. The price level attainable for this commodity is curtailed by the costs of production of synthetic rubber, whose development resulted partly from an earlier attempt to manipulate the price of the natural product.[18] The cocoa agreements, the first of which was the product of ten years of negotiation, were ineffective until 1981 because market prices were usually above the agreed price range and the resources of the buffer stock consisted in cash, not cocoa. The Ivory Coast (the largest producing country) and the USA (the largest consumer) declined to join a new agreement made in 1981 – the former on the ground that the agreed price range was too low, the latter on the ground that it was too high. The buffer stock then began buying cocoa to arrest a decline in prices, but by April 1982 it had exhausted its cash, accumulated a stock of 100,000 tonnes, and still failed to bring market prices above the agreed minimum.

The coffee agreements, which build on a long history of Brazilian attempts to valorize the commodity, were intended from the beginning to raise coffee prices, not merely to stabilize them about a market trend. The means to this end are export quotas allocated among countries in proportion to their productive capacity. There is no international buffer stock; an excess of production over quota in any country is meant to be neutralized by local stocking. The limitation of sales may be assumed to have raised aggregate proceeds, but against this gain must be set the costs of producing unsaleable coffee; by 1982, stocks in the producing countries were equal to about one year's world consumption. The allocation of quotas among the producing countries has naturally been strongly contested, and there have been frequent complaints by the Ivory Coast and other African producers that Brazil and Colombia enjoy excessive market shares. It is likely that not all producing countries have been advantaged by the increase in aggregate coffee earnings, and that the Ivory Coast in particular – as a large and low-cost producer mainly of the *robusta* variety used for soluble coffee – would have fared better in an uncontrolled market.

So far, international commodity agreements have contributed little toward raising the export earnings of West African countries and, more particularly, the revenues of their governments. As with many other achievements of economic diplomacy, the attention given to negotiation of these agreements appears incommensurate with their practical importance. Attempts to find more assured and less restrictive forms of external subsidization will no doubt continue, but experience provides little basis for a belief that this aim can be realized by multilateral agreements to manage the prices of commodities in world trade.

Deficit financing

Relatively large fiscal deficits appear in Table 22 for The Gambia, Ghana, Liberia, Mauritania, Nigeria, Sierra Leone and Togo. In some cases, like Liberia, they reflected largely accumulation of external debt. In others, deficits were internally financed – by drawing on official reserves, by borrowing voluntary savings (usually through financial intermediaries), by obtaining from the banking system credit created specifically to meet the government's needs, or by some combination of these methods.

Deficits in public finance are scarcely avoidable. They may result from variation over time in the planned expenditures of a government – a variation which could well be pronounced in economically small countries. Financing a deficit in one year by drawing on reserves retained from earlier years' revenues, or by long-term borrowing made in anticipation of higher revenues in future years, can be regarded as means of equalizing over time the flows of public receipts and outlays. Alternatively, a deficit can result from mistaken budgetary forecasts. Divergences between revenue estimates and outturns are likely in West African countries, where receipts depend so heavily on export prices, the profitability of a narrow range of enterprises, or the volume of imports. In addition, expenditures may exceed the forecasts because of cost overruns in capital works, unanticipated wage increases, higher import prices, or simply failure to control the spending of the constituent parts of an administration.[19] Involuntary deficits arising in these ways are often covered by short-term borrowing. It would seem prudent that they should be followed by more conservative estimates of revenue or a more stringent regulation of spending. But the pressures responsible for the initial deficit – falling export proceeds, rising import costs, administrative incoherence, or whatever – are not necessarily short-lived, and the first deficit may therefore be followed by others.

In practice, the boundary between involuntary and intentional resort to short-term borrowing may be indistinct. The domestic banking system has peculiar attractions as a source of government finance. The competition of alternative outlets for the funds deposited with commercial banks can be checked by exchange control and by increasing the cash ratio (i.e. the proportion of their deposits held as currency and deposits with the central bank) that the commercial banks are obliged to observe. The commercial banks can also be obliged to purchase quotas of government bills, or the central bank itself (so long as it is under exclusive national control) can become a practically unlimited spring of credit, the legislation limiting its power to lend to the government being amended as necessary, if not simply ignored. Once a country has acquired an autonomous monetary system, the facility with which bank credit can be obtained may produce a persistent tendency for government spending to exceed the proceeds of taxation, grants and long-term borrowing.

This tendency is likely to have inflationary consequences. They may be countervailed if the price-elasticities of supply of goods and services are high, if the ratio of commercialized to total economic transactions is rising, or if the demand to hold money balances is increasing. These factors possibly offsetting the effect on prices of monetary expansion are unlikely to be very durable in West Africa. Sooner rather than later, recourse to financing government deficits by creation of credit will drive up prices and depreciate the national currency. The outcome has been likened to a tax on the holding of money. Money balances must be increased if their real value is to be maintained. Holders of money therefore reduce their use of goods and services. The resources thus freed then become available to those who have access to the bank credit which has produced the inflation – either the government alone, or the government along with private borrowers. Like other taxes, this 'inflation tax' shifts purchasing power toward the government, or toward interests the government wishes to favour. On this ground, deliberate recourse to deficit financing has sometimes been defended as an instrument of policy – a kind of auxiliary tax system to which a government may legitimately resort when its revenues more conventionally obtained fall short of national needs.

Those who are subject to the inflation tax try to avoid or offset it, as they do other taxes. They may economize on money balances, thus resisting the attempt to reduce their use of real resources. They may attempt to pass on the tax by marking up the prices of the goods and services they offer for sale, or by claiming higher money-wages or allowances. They may even avoid the tax by withdrawing partially from the monetized economy. In so far as these defensive stratagems are effective, they propagate the inflation; they increase the monetary demand for goods and services relatively to the supply (or the supply of money relatively to the demand to hold it). Ironically, the government itself may become foremost among the economic actors seeking protection from inflation. If the tax system is of low elasticity with respect to increases in domestic money-income (as is likely to be the case in West Africa), government revenues will tend to rise less rapidly than prices, and therefore less rapidly than government expenditure. Resort to credit creation to fill the government deficit in one year may thus itself be the cause of a deficit in another year. The defence available to the government against this consequence is, of course, to borrow further from the banking system. Here, then, is another force tending to make deficit financing a chronic tendency rather than an occasional expedient in countries possessed of autonomous monetary systems.

Ghana is the leading illustration in West Africa of intentional use of the inflation tax, and of the propensity of that tax to grow with use. A first phase of deficit financing was in the period 1961–6; the money supply was then expanded at a compound annual average rate of 11 per cent, and prices rose by 13 per cent on annual average according to the official consumer price index. A second and more explosive phase began in 1974. Between 1973 and 1981, the compound annual growth rates are recorded as 42 per cent in the money supply and 61 per cent in consumer prices; in two years (1977 and 1981), prices more than doubled according to the official index. What had begun in 1961 as a mechanism, ancillary to orthodox taxation, of securing resources for official use had become, by the late 1970s, a desperate struggle by government to keep its spending power in line with rising prices.

Using resources for official purposes

Data on government expenditure, or the uses made of the resources obtained for official purposes, are usually analysed by function and by economic type. The functions of government expenditure usually identified are general administration, defence, social services and economic services; the social services are subdivided as education, health, social security and welfare, housing, and the like, and the economic services are allocated among sectors such as agriculture and transport. The economic types of government spending are current expenditure, capital expenditure and lending; current expenditure is further divided as wage and salary payments, other purchases of goods and services, interest payments, and subsidies and other current transfers.

Both kinds of analyses have appeared in the *Government Finance Statistics Yearbook* of the IMF for years in the 1970s, and international comparisons appear to be possible,[20] but the data are deficient in important ways. Not all the West African countries are covered. The outlays of parastatal bodies, including public marketing organizations, are included only to the extent that they are subsidized by the central government. Expenditures by tiers of government other than the central or national government are also excluded from these analyses of the distribution of public expenditure; this omission is particularly important in Nigeria, where the State and local governments were spending about 25 per cent of centrally collected revenues in the late 1970s,[21] in addition to independent revenues of their own. Some expenditures of aid, notably the use made of technical assistance, are also omitted. Finally, because of recording deficiencies, the sum of expenditures classified by function or economic type fails frequently to agree with the total whose distribution is being analysed.

The findings that follow must therefore be regarded as tentative and of uncertain significance. It appears that general administration (legislative bodies, police, justice, tax collection, foreign affairs and other overheads) took between one-fifth and one-quarter of government spending in most West African countries about 1978, but that this proportion was appreciably lower in federal spending in Nigeria. The share of defence approached one-fifth in Nigeria and, more surprisingly, in Guinea-Bissau, Mali and Upper Volta, but was usually less than one-tenth.[22] Social services mostly took one-quarter to one-third of total expenditure, but their share was much higher (nearly one-half) in the Ivory Coast. Education was much the most important social service, often accounting for 15 to 20 per cent of total expenditure, and in the Ivory Coast for about one-third. The share of health was usually 6 to 8 per cent. Economic services obtained anything between one-seventh of total expenditure (Mali and Senegal) and nearly one-half (Nigerian federal spending).

Current expenditure comprised less than one-half of total expenditure about 1978 in The Gambia, Liberia and Nigeria, but was about two-thirds in Togo and four-fifths in Ghana, Mali, Niger, Senegal, Sierra Leone and Upper Volta. Net lending[23] was important in Nigeria, where it constituted one-fifth of federal spending, mostly as loans to the State governments. Not surprisingly, the share of establishment costs (wage and salary payments) tended to vary directly with the share of current expenditure; it was low (15 to 20 per cent) in The Gambia and Liberia, and high (40 to 50 per cent) in Mali and Senegal. Interest payments were highest (14 per cent) in Ghana.

Probably the most important determinant of patterns of government expenditure at these levels of aggregation is the rate at which government receipts (in real terms) are growing. Simply because capital expenditure is more flexible than recurrent, a rapid acceleration in receipts tends to increase the relative importance of the former. In turn, a relative shift toward capital spending is likely to reduce the relative importance of general administration and defence and to increase that of social and (perhaps especially) economic services; such changes can be observed in Nigeria during the 1970s.[24] A greater proportion of capital spending will also depress the relative size of establishment costs. Conversely, when receipts grow sluggishly or not at all, general administration is likely to bulk large in the functional analysis of government expenditure, and establishment costs in the analysis by economic type. If this stationary phase succeeds one in which receipts were increased by large short-dated borrowings, interest payments may also assume considerable relative weight. Governments short of revenue and suffering the rigours of an IMF stabilization programme have found themselves with spending capacity largely pledged to their least flexible commitments, namely paying their employees' remuneration and servicing their debts; Liberia in the mid-1960s was the first West African experience of this condition.

The distribution of public expenditure does not, of course, merely react to changes in the availability of resources. It also reflects policy objectives. Increases in capital spending, for instance, have been supported by the belief that raising the investment rate is a necessary (or even a sufficient) condition of economic growth (see Chapter 6). Spending on economic services is seen as a means of enlarging productive capacity, spending on social services as a means of delivering welfare. (The belief that public spending on capital formation and economic and social services is more effective than private spending in raising productivity or welfare becomes a justification of attempts to increase the ratio of government expenditure to GDP.)

The distribution is more finely influenced by judgements concerning the relative yields of alternative outlays in economic output, social welfare or the satisfaction of other national needs such as security or international stature. Not only are choices among governmental functions implied. In addition, many public expenditures are specific as to place, and a location (of, for example, a manufacturing plant, a hospital or a public housing project) must be selected. Public money can also be spent in alternative institutional ways, among which choices must be made. For example, an industry may be publicly supported by the provision of free professional services by a government department or agency; through the subsidization of some of its inputs such as electricity or agricultural raw materials; by tariff protection or fiscal concessions (implying a sacrifice of government revenue rather than government expenditure); in contributions to capital funding, as loans or equity participation; by complete public ownership and control; or in some combination of these methods. Again, the welfare of urban workers may be raised by increases in wages, or by subsidizing basic foodstuffs, or by expenditures on sanitation, electricity supply, housing and schooling.

These choices among the functions, locations and institutional forms of public spending are not made (and, in practice, cannot be made) by persons who are themselves detached from and uninterested in the outcomes. Public expenditure decisions inevitably become related to the maintenance of political control, the promotion of party interests, and the ambitions of individual statesmen. Chief executives and legislators in independent African states have been prone to confuse national needs with their personal interests; an early perception of this tendency in the francophone states led one commentator to draw an analogy with the court of Louis XVI.[25] Where electoral politics have been practised, public funds have been often used with a view to electoral advantage. Even governments not at a risk in elections must maintain their acceptability to politically influential groups.

These relationships between national objectives and narrower interests, this absorption of policies by politics, will be further discussed in Chapter 6. The point that requires making in the present context is that because special interests (of politicians, government departments, parastatals, communities, occupational groups, and private businesses) do become vested in public expenditures, it is often difficult to make more than marginal changes in the pattern of spending at any time. Choices of where and how and on what to spend cannot be made without regard for what is already being done. Every decision on these matters is to some extent constrained by previous decisions. Government borrowing can be regarded as an attempt to relax this constraint, as well as actually to increase the volume of resources at public disposal. Only in the event of some rapid surge in public receipts, accompanied perhaps by a large displacement of the ratio of government expenditure to GDP, does a considerable degree of freedom appear in the allocation of public resources. Such conditions did obtain in West Africa in the late 1940s and the 1950s, because of buoyant tax receipts from external trade, marketing board surpluses, and budgetary support provided by France and Great Britain. They persisted in the Ivory Coast, at least until 1979. In Nigeria, they reappeared in 1974–80 with the oil booms. Elsewhere, these conditions have been contrived through short-dated external borrowing and deficit financing, but with only transitory success.

The distribution of public spending is also strongly influenced by practicability or relative costs. Obviously, costs are weighed against anticipated benefits when choices are made. These comparisons have an important bearing on choice of location; thus the 'urban bias' frequently alleged against government expenditure is partly attributable to the cheapness of providing some services to large concentrations of populations as compared with small communities. Again, relative costs might appear sufficient ground in African countries for choosing to deliver welfare through public amenities and services instead of through paying cash benefits to individuals. In addition, practicability can continue to affect the distribution of public expenditure (more particularly, the functional distribution of capital expenditure) after the choices among alternatives have been made. Actual expenditures under development plans frequently fall short of those intended, even when finance is not a constraint. The explanations include inadequate project preparation, lack of know-how, shortage of executive capacity, and difficulties in material procurement and importation. The incidence of these factors is not uniform among sectors. Underspending tends to be less on purposes which are already familiar, for which there are established models that can be replicated, and where supplies of material and technical expertise can be readily obtained. Where these conditions are exceptionally favourable, spending may even overshoot targets. Conversely, spending tends to lag markedly on purposes with which there is little local experience, and whose profitability or benefits may therefore need to be argued to the satisfaction of foreign partners or aid donors. Spending 'slippages', as they have been called, therefore tend to be less marked in general administration, defence, public works and schooling, and more pronounced in economic services involving technically sophisticated forms of infrastructure and manufacturing.[26] In practice, therefore, certain functions may expand in relative importance simply because they are relatively easy (or administratively cheap) ways of spending money.

Influencing private use of resources

Intentionally or inadvertently, the use made of resources left in private hands is influenced both by taxation and by government spending. Since no tax system is perfectly uniform in its incidence (and those of West Africa are conspicuously lacking in that attribute), taxation tends to divert resources from penalized activities to those that are relatively or absolutely exempt. It might be expected that in West Africa the diversions would be away from exporting activities, the trade in import goods (especially finished consumer goods, on which tax rates are heavier), and formalized enterprises and occupations. These tendencies need not appear in practice. They can be countervailed by inadequate administration (allowing much tax evasion) or by blatant illegality (as in the smuggling of imports and exports). They will remain only latent if, in spite of taxation, the comparative advantage for farmers in growing crops for export is still intact, if the elasticities of demand for imports remain low, and if advantages survive in the formalization of business practices and employment. The impact of uneven or discriminatory taxation on structures of production and demand therefore depends on the rate of tax and the efficacy with which the tax is levied. It appears that taxation of agricultural exports in West Africa (imposed in various ways, as was explained earlier in this chapter) has been such as actually to deflect labour and land from those forms of cash-cropping. For the most part, this result has been inadvertent and is officially deplored.[27] On the other hand, the reduction of consumer imports by taxation, while often intended in order to shift demand in favour of domestic production, has been difficult to accomplish, and the tariffs have had to be reinforced by import licensing in some countries. While high comparative disadvantage in import-substitution partly explains this difficulty, currency overvaluations have also been instrumental in Ghana, Nigeria and Sierra Leone.

Since transfer payments and subsidies are, in effect, negative taxes, they produce tendencies opposite to those of taxation; they tend to attract resources to favoured spenders or activities. Thus, subsidies have increased the demand in West African countries for electricity, schooling, water, imported foodstuffs, fertilizers, pesticides, petroleum products and capital equipment. Administrative shortcomings may restrict these effects, as they do those of taxation; in particular, tax rebates appear to be a practically ineffective method of subsidization.[28] A further limitation may be imposed by the fixing of an amount available for payment of a subsidy; once the vote is exhausted, the subsidy terminates – and so, perhaps, does the supply of the product that was subsidized. This eventuality brings out a feature of many subsidies: uncertainty concerning who benefits (or even concerning who is intended to benefit). If, for example, fertilizer or a foodstuff is being imported only through subsidized channels and some demand is left unsatisfied, the available supply may find its way into free and illegal markets where it is resold at market-determined prices. The beneficiaries are then the importers or distributors, not farmers or consumers.

Some subsidies are implicit, for government expenditures that involve direct public use of goods and services also influence private uses through their effects on the relative costs and efficacy of alternative current outlays, investments and technologies. Improvements in transport and communications are an obvious illustration of public expenditures that affect private costs. Flood control, irrigation works and reforestation may exercise profound influence on production possibilities. Expenditures on education, sanitation and preventive medicine are scarcely less germane. As these last examples suggest, consumers as well as producers have interests in how (and where) the government is using resources.

Some implicit subsidies result from implicit taxes. In these cases, no revenue is yielded, but resources are nevertheless transferred from penalized to favoured groups or industries. Currency overvaluation has already been mentioned as constituting such a nexus. Producers of export goods are implicitly taxed since they receive less in local currency than the goods would be worth if their foreign exchange value were converted at a market rate of exchange. An equivalent implicit subsidy is then received by traders in or users of import goods, and other persons having access to foreign exchange, since they pay less for this resource than it would cost at a market rate of exchange with the local currency.

Analogous cases arise from official determination of other prices. Suppose that a government attempts to hold down urban living costs by stipulating low prices for locally produced foodgrains, as has often been the practice in the West African Sahel.[29] Urban consumers are then subsidized not from government revenues but effectively from the earnings of farmers, which are lower than they would be in a free market. Official regulation of the interest rates charged on bank lending provides another illustration. The maximum rates allowable are often fixed at levels which, taking account of inflation, are very low or even negative in real terms.[30] The rates paid by the banks on interest-bearing deposits are consequently depressed, and savers are being implicitly taxed for the benefit of borrowers.

In practice, the bulk of locally produced foodgrains is likely to be sold outside formal channels, perhaps unlawfully, and at market-determined prices. The consequence is that the incidence of the implicit subsidy is narrowed; the beneficiaries are those who have access to the official, price-controlled supplies and who do not have to resort to the free market as buyers (if they appear in that market as sellers, they are collecting the subsidy in cash instead of in kind). Free informal markets for foreign exchange and for credit may similarly parallel the controlled official markets, with some users having to meet market conditions while others are privileged.

Licensing

The purpose of licensing may be to raise a revenue, as with motor vehicle duties in West African countries. In such cases, licences are sold without limitation. Where licensing is restricted or discretionary, an attempt is being made directly to control resource use, and often to control the distribution of an explicit or implicit subsidy. Restrictive licensing is complete when no licences are issued, or in other words when an activity is forbidden, as the import or export of a particular product may be forbidden. So, where licensing has a restrictive purpose, a licence can be regarded as exemption from a prohibition, a right to do what others cannot lawfully do.

Fiscal concessions offered as inducements to foreign enterprises are contingent on the licensing of establishment. Trade in a product that has been cheapened by currency overvaluation or by purchase at official prices will be licensed. In such cases, licensing can be regarded as following in the wake of subsidy; it is an attempt to ensure that something of value (in terms of public policy objectives) is obtained in return for the benefit that has been made available. These cases also show a tendency for some indirect influences on resource use to become absorbed in direct controls.

The licensing of industrial establishment will depend on a would-be investor demonstrating a need (other than the profitability of his own capital) for the investment he proposes. He may argue the creation of wage-employment, and of outlets for investible funds other than his own; the saving of foreign exchange by substitution of his production for imports; the introduction and adaptation to local conditions of an unfamiliar technology; the public revenues that could be obtained from exploiting natural resources presently unused; or the contribution which his investment could make, directly and indirectly, to economic diversification and stability. For their part, the licensing authorities may make issue of a licence conditional on the nature of the products to be produced, the location of the investment, the number of jobs to be created (or an acceptable ratio of labour to capital), the participation of other investors (indigenous as well as foreign, or public as well as private), agreement on programmes to reduce over time the import-coefficient in the production and to increase the inputs of domestically produced raw materials and semi-products, determination of a particular tax regime where natural resources are being exploited, or commitments by the investor to employ national rather than foreign personnel, to train labour and perhaps not unduly to restrict the dissemination of his technology.

A licence to produce is therefore the outcome of a bargaining process. The applicant represents the potential benefits for interests other than his own, or more particularly for interests he believes that the government wishes to favour. The government offers not merely a right of establishment but a right not freely available, or in other words a privileged status in production; and it may offer such further support as a protective tariff, tax exemptions, low-cost credit or risk-reducing contributions to capital. The bargaining is predicated on the beliefs that the government ought to secure gains, additional to the private returns on the investment proposed, for the national economy it manages, and that the government in return ought (or at any rate can be induced) to restrain competition with and perhaps also fiscally to subsidize the investment. Industrial licensing would not disappear in the absence of these beliefs, but its scope would be greatly reduced.[31]

Either party in this bargaining process may overplay his hand. A prospective investor may ask for concessions that are judged incommensurate with the benefits he will confer; a less intransigent applicant may then be preferred. Alternatively, the government may propose conditions that appear unreasonable to the investor; or it may confront him with procedures of authorization so cumbersome, dilatory and uncertain in outcome as to be a sufficient deterrent.[32] If the investor is foreign, he may then look elsewhere for rights of establishment; if he is indigenous, he may revert to those informal economic activities where a right of establishment is

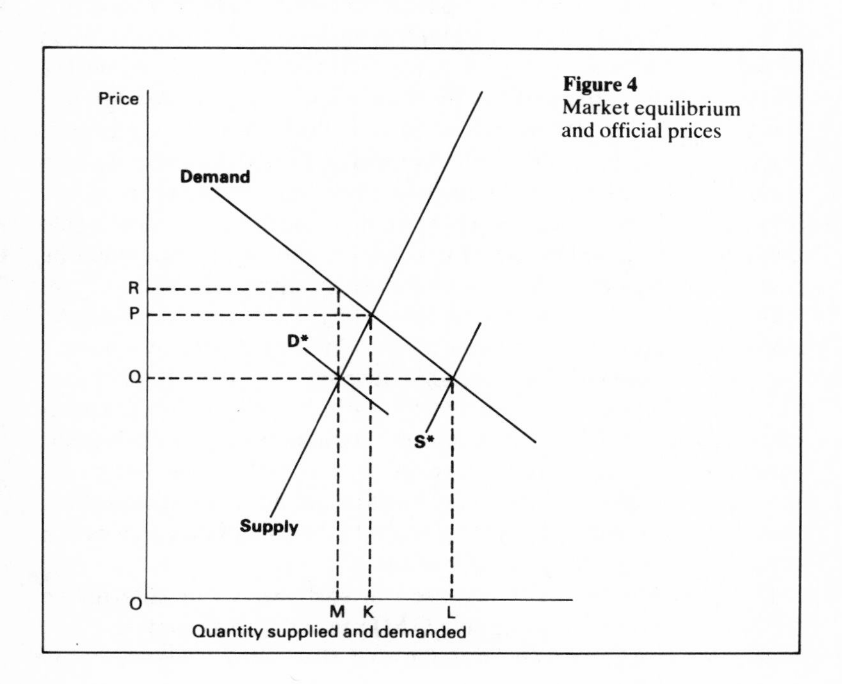

Figure 4
Market equilibrium and official prices

neither necessary nor obtainable.

Like licences to produce, licences to trade can represent two kinds of privilege. First, licensing may act merely to restrict competition, sometimes for the benefit of established businesses, sometimes for that of new businesses (such as a state corporation in the import trade) that the government wishes to promote. Second, if the price at which licensed traders may buy is officially fixed below the free-market equilibrium value, the licence to trade can become also a licence to share in an explicit subsidy.

Thus, in Figure 4 the demand for a good is shown negatively related with its price (the higher the price, the less buyers will want to buy), and the supply positively related (the higher the price, the more sellers will want to sell). The good may be foreign exchange, credit, an export crop like cocoa, imported rice or wheat, fertilizer or seeds, locally produced millet or maize. The price OP is the market-clearing equilibrium at which the quantities demanded and supplied are equal at OK. The government might license purchases at this price OP. The consequence would be to limit competition for the good; buyers who would have been willing to pay more than OP in order to secure access to supplies will not be able to do so.

Suppose that through official intervention (concerning the value of foreign exchange in terms of the domestic currency, allowable rates of interest, the ex-farm prices of export crops or foodgrains, the ex-factory or import prices of agricultural inputs, the import prices of staple foods) the price is fixed at OQ, below the market-clearing equilibrium. A disjunction now appears between supply and demand. At the price OQ, supply is reduced to OM while demand is increased to OL. In principle, this disjunction might be removed by a publicly financed increase in supply (shown in Figure 4 by a shifting of the supply schedule to S*). Thus, reserves might be drawn on, external borrowings made or more aid solicited to meet the excess demand for foreign exchange created by an overvaluation of the domestic currency or to pay for additional imports of staple foodstuffs or agricultural inputs; but, obviously, this solution is practically limited. Again, the central bank might expand the supply of money to meet the excess demand for credit resulting from low allowable rates of interest, but inflation, or an acceleration of inflation, is the likely consequence.

If supply is not to be increased in such ways, the alternative solution is to ration the supply (OM) presently forthcoming at the official price by licensing purchase of the supply. The quantity demanded is then matched with the available supply. But a disjunction remains between the official price and the price that the available supply would command in a free market. The latter is represented in the diagram by OR. QR is therefore the value, over and above the official price, that the good commands, given the restriction of supply to OM. This value is potentially available to a licensee who chooses to resell the quantity (of foreign exchange, credit, cocoa, imported rice, fertilizer, millet) that he has been allocated. So far as licensees realize this potential value, they are encashing the subsidy provided by official policy; the subsidy has then become an administratively generated rent.[33]

To enrich licensees is not usually the purpose of public policy, or at any rate not the manifest purpose. Official prices are held below market equilibria in order to keep down the cost of living, to encourage production of specific goods (especially manufactures and staple foodstuffs) by reducing their production costs, to increase investible surpluses, and to produce uncommitted public revenues. The private appropriation of monopoly profits or administratively generated rents is therefore likely to be resisted. In industrial licensing, as has already been observed, some *quid pro quo* will be demanded of the licensee. In the licensing of trade, attempts will be made to ensure that the benefits accrue where the government intends.

The latter task is not an easy one. Producers who are offered low official prices have a powerful incentive to circumvent licensed traders in favour of illicit buyers who pay a price nearer the market value of the produce. Consumers whose demand cannot be satisfied at official prices because of restricted supply, and who possess the means to buy at free-market prices, may be no less ready to pay the full value of a good. Licensed traders will wish to realize the administratively generated rent that has been placed in their hands. Consumers whose demand is met at official prices may become traders themselves and sell at higher prices to those whose demand has not been met. Alternative or parallel markets thus grow up alongside those officially sanctioned. As a further twist in economic relationships, the public officials responsible for issuing licences, or for keeping trade within official channels, may appropriate part of the unearned income that official price policy has created; they may do this individually, by accepting bribes in return for licences or for deflecting goods out of official distributive channels, or collectively, by inflating the costs of the organizations they staff. In short, many actors in the economy will be striving either to beat official pricing or to profit from it.

Monopolization of export-crop sales illustrates some aspects of this struggle to exploit or escape government intervention in pricing. It has been recognized since the 1960s that the primary purpose of these marketing monopolies is to raise public revenue. This purpose can be partially frustrated by the generosity of the margins allowed to the licensed buying agents[34] (though both the selection of the agents and the generosity of their allowances may reflect other political objectives). Perhaps more importantly, the marketing board itself may become a costly system of administration, with high overheads, a padded payroll, and numerous semi-charitable obligations;[35] to this extent, while public revenues are raised by the monopoly, they are committed to or absorbed by the costs of tax collection. Finance for crop purchase is extended to its agents by the board or by the banks; if more profitable uses than crop purchase are found for it, buying agents appear who can pay farmers only in chits or promissory notes.[36] Farmers are faced by a price that is usually low in relation to the market value of their crop, payment possibly in chits whose redemption is uncertain, and an absence of alternative marketing outlets within the law.

One solution is to market outside the law. Either the farmer, or a trader to whom he sells illicitly, smuggles the produce to an adjacent country where official buyers are paying a better price.[37] This better price may actually be higher (if the foreign currency in which it is paid were converted at the official rate of exchange), or it may be better simply because it is paid in foreign exchange, another good which may have a free-market value markedly more than its official value if access to it is licensed. Where access to foreign exchange is controlled, the proceeds from smuggled exports will be used, directly or indirectly, to pay for smuggled imports and foreign services and to make transfers and investments abroad.

Faced by this competition for appropriating the excess in the marketable value of a crop over its official value, the government may press for greater efficiency in the marketing board and firmer constraints on the operations of the licensed buying agents. It is likely to experience the resistance of strongly entrenched interests –interests which it may well have created itself to serve political ends. Farmers and the illicit traders may appear more vulnerable to government pressure. Official producer prices may be depressed even further. Smuggling will be represented as among the most heinous of crimes, and the penalties increased. The results may be disappointing. Preventing smuggling is the responsibility of the police and border guards. They too have opportunities for sharing in the administratively generated rents. They too can be in the business of issuing exemptions from prohibition.

Consider next the goods intended for home markets whose prices are subsidized. Subsidization may be explicit (through public spending or tax remission to cover part of production or import costs) or implicit (through the official determination of input prices, rates of interest, or the exchange rate). The goods may be home-produced or imported; home-produced goods with a substantial import content fall into both categories in the sense that they, like imports, can be subsidized by an overvalued currency.

Assume that final users (consumers in the case of foodstuffs, farmers in that of fertilizers, investors in that of loans) are intended to be the beneficiaries of these subsidies. How can this result be ensured? How can producers be prevented from circumventing official controls, and distributors from benefiting from those controls? In principle, a solution is to extend licensing throughout the distributive chain to include all final users. If all purchases, including those of final users, were conditional on the surrender of ration coupons, the demand unsatisfied at official prices could be eliminated (as is shown in Figure 4 by a shifting of the demand schedule to D*); there would be no distributive channels alternative to the officially sanctioned channels, and only final users would benefit from the subsidy. In practice, this solution is not generally available in West Africa. Not only does it make extreme demands on administrative resources (including probity) but also it requires effective control of the borders – otherwise supplies will be diverted from the licensed users at home to the unlicensed users in neighbouring countries whose demand is not restrained by rationing.[38]

Public enterprise seems often to be regarded as an alternative means of preventing the embezzlement of subsidies. Where imports are licensed, a public corporation can be set up to participate in, or even to monopolize, the import trade. Where imports of staple foodstuffs are subsidized, another corporation can be made responsible for their distribution. Where agricultural inputs (seeds, fertilizer, pesticides, tractor services) are made cheaply available to farmers, their supply can be entrusted to a ministry of agriculture. Where the rentals of urban housing are to be held down, a government housing corporation can be made responsible for new construction.[39]

These are imperfect solutions. Take the case of public credit institutions. Interest-rate controls produce a disjunction between the supply of and the demand for loanable funds in the banks. If the banks are left to apply their own criteria for allocating the available supply, they will concentrate lending in directions where their administrative costs are low, this being the only means available for maintaining profitability. They will lend to large-scale and well-established borrowers able to provide acceptable collateral. This tendency could well be inconsistent with official objectives – if, for example, the government seeks to secure a preference in lending for agriculture as against commerce, or for indigenous enterprises as against foreign.

Specialized credit institutions may therefore be set up in the public sector, charged with catering for the needs of farmers, indigenous businessmen, or other groups deemed to be worthy of low-cost financial support. One way of financing these institutions is to increase the reserve requirements of the deposit banks. Loanable funds are thus diverted from the deposit banks to the central bank and can be used by the latter as working capital for the specialized institutions. It becomes even more necessary for the deposit banks to find low-cost and low-risk outlets for the funds left with them. The public lending institutions can apply other criteria of allocation. But in neither case are the available loanable funds channelled toward the uses in which they will be most productive. Cheap credit is almost bound to be used in an economically undiscriminating way. Public credit, if West African experience is anything to go by, is almost bound to be extended with political aims in view.[40]

The banks cannot hope to mobilize all savings at the low interest rates they have to offer depositors. Savings are largely channelled to informal credit markets, where they meet with would-be borrowers who have been repelled from the banks and the specialized credit institutions by the application of criteria of allocation. The supply price of the savings will be higher in these markets because the risk of default is much greater in informal credit agencies than in banks whose liquidity is regulated by law. Borrowers in these markets pay not only more than the banks do charge but also more than they would charge if their lending rates were not controlled. The costs of subsidizing credit are therefore borne by those borrowers excluded from the officially-regulated sources of credit as well as by those savers who do deposit with the banks. Further, since the informal markets are localized and in other ways segmented, there is likely to be a loss in the efficiency with which loanable funds are used, on top of whatever losses are incurred through the application of allocation criteria by the banks and the specialized credit institutions.[41]

Similar stories could be told of other subsidized goods. So long as demand is unsatisfied, either at home or in neighbouring countries, and the good is transferable, a public marketing institution is unlikely to maintain control of the supply. Supplies will be diverted into parallel 'black' markets, even if they get as far as final users. Far from being the solution of how to distribute subsidized goods equitably or efficiently, the public institution can well become part of the problem. It is likely to have high operating costs, partly because it is not exposed to competition in the business entrusted to it and partly because it relies on direct methods of allocating supplies instead of on prices; hence it will tend to absorb the benefit intended for final users. And, where prices do not allocate supplies, political criteria are apt to take their place. The public institution wields patronage; those who have use for the subsidized good become clients competing for its favour.

The politicization of distributive trade becomes more apparent when responsibility for allocating supplies is assigned to branches of the ruling party or to the constituent parts of a corporate state. In Ghana in 1981, for example, the supplies of fifteen 'essential commodities' in excess demand were entrusted for distribution among consumers to such bodies as the Trade Union Congress, the Civil Servants' Association, the National Association of Teachers, the police, the universities, the Farmers' Council, the Cocoa Marketing Board, the Food Distribution Corporation, and the ministries of defence, education and health. Each of these bodies was to provision its own personnel and clientele as it saw fit. Allegedly, only 8 per cent of the available supplies were left for those consumers (probably constituting the majority of the population) who lacked connections with, and the protection of, any of the approved bodies corporate.[42]

Price controls at the retail level are also favoured as a supposed means of ensuring that subsidies go where they are intended. Rationing of final users, were it feasible, could be reinforced by such price controls, but the controls are not an alternative to comprehensive rationing; in its absence, they are only an attempt to substitute a legal restraint for the economic restraint that is needed, and they will be either ineffective or effective only in creating a privileged group of users. Thus, interest-rate controls, as has already been argued, cheapen credit for borrowers who have access to formal credit institutions and make it more expensive for those who do not. Similarly, price controls on consumer sales are likely to benefit customers who buy in the large urban shops, where price surveillance can be effective and where distributive costs may be lowest, but actually to disadvantage customers who buy in the markets and in rural areas.[43] As with smuggling, so with retail price controls, official recognition of failure engenders much political rhetoric and efforts to intimidate traders.[44]

Summary

Governments attempt to implement economic policies partly by securing resources for their own uses and partly by influencing, directly or indirectly, the uses to which resources left in private hands are put. The ratio of central government expenditure to GDP is a rough (but incomplete) indication of the extent of public resource use; in West Africa, it appears to range between 15 and 40 per cent, with high ratios found in the larger and better-off countries (Nigeria and the Ivory Coast).

Resources are secured by taxation, earnings from property, foreign aid and borrowing. Tax structures are heavily biased toward levies on foreign trade, or on the operations of industries producing largely or entirely for export. The explanations are the facility with which import and export duties can be collected, and the high proportions of economic rent in the value of some mineral exports, especially oil.

Taxation of agricultural exports developed out of statutory purchasing monopolies, or marketing boards, established in British West Africa in the 1940s, supposedly for the purpose of economic stabilization. Their avowed purpose was beyond the capacity of the boards actually to achieve, and they became, in practice, fiscal instruments of government, persistently paying to farmers much less than the export value of their crops. The marketing of export crops evolved in a similar way in the francophone countries, although there it was not until the 1960s – or, in some cases, the 1970s – that farmers were heavily squeezed to provide government revenue. The proportions of farmers' earnings withheld by overt and covert taxation have apparently ranged, among various countries and various periods of time, between about 20 and 60 per cent –without allowance for the negative effects on those earnings of inefficiency in the statutory marketing systems and of currency overvaluations in some countries.

The bulk of foreign aid does not appear in government accounts, partly because it is received in kind, not in cash. In the late 1970s, aid to West Africa was running at an annual rate of $1.8 billion: two-thirds as grants, one-third as loans with a grant element usually exceeding 50 per cent. For the West African region as a whole, aid does not appear of great relative importance – about 2 per cent of GNP, 6 per cent of investment and 10 per cent of imports in 1979. If Nigeria, to which aid was negligible, and the Ivory Coast, where aid was of much less than average importance, are excluded, the relative magnitudes for the remaining countries are much greater – about 9 per cent of GNP, 41 per cent of investment and 35 per cent of imports. Strong aid relationships result from attempts to discharge what are considered to be the appropriate responsibilities of national government in countries which are economically too small to sustain this task unaided.

Those relationships would be less irksome to the beneficiary governments if the aid were obtained as of right, instead of at the discretion of donors. Thus, aid might be obtained through trade, if export prices were determined administratively instead of by market forces. International commodity agreements have come to be regarded in this light by West African governments, but their contribution to increases in the revenues of those governments has so far probably been small.

Government borrowing may be made externally, internally from voluntary savings, or internally by credit creation which, if it results in inflation, has effects analogous to a tax on money balances. Resort to bank finance is scarcely avoidable, since deficits in government accounts are likely to appear from time to time involuntarily. But the ease with which credit can be obtained by government, where an autonomous monetary system has been instituted, can produce a persistent tendency for public spending to exceed the proceeds of taxation, aid and long-term borrowing. Inflation is propagated by the efforts of holders of money balances to find protection from or compensation for rising prices. If the tax system is of low elasticity, the government itself may be foremost in this struggle; Ghana since 1974 has been a conspicuous illustration.

The uses to which public resources are put are analysed by

function (e.g. defence, education) and by economic type (e.g. establishment costs, other current expenditures, capital expenditures). The results of these analyses depend partly on recent rates of change in public receipts, and therefore differ appreciably among countries. The pattern of public spending is also influenced by policy objectives in which particularist interests become embedded, and accordingly this pattern may be difficult to change except when receipts are buoyant. Practicability also affects the distribution of government expenditure, especially capital spending; some functions may expand in relative importance simply because they are relatively easy (or administratively cheap) ways of spending money.

The use of resources left in private hands is influenced, intentionally or inadvertently, by both taxation and government spending. Resources tend to be repelled by taxation and attracted by subsidization. Some subsidies are conferred implicitly by public spending on objects like road building, flood control and preventive medicine. Some implicit subsidies result from implicit taxes; no revenue is generated, but resources are nevertheless transferred from penalized to favoured groups or industries. Official price determination (of foreign exchange, agricultural products, credit) has these consequences, though they are likely to be partly frustrated by the development of black markets for the affected goods or services.

Where the official licensing of production or trade has a restrictive purpose, it is an attempt to ensure that policy objectives are served by the privilege which has been granted (and which may include explicit or implicit subsidization). In industrial licensing, the government trades protection against competition, and possibly tax concessions, for such prospective gains as employment creation and savings of foreign exchange. The licensing of distributive trade results often from a need directly to allocate supplies, when a price is fixed officially below the free-market equilibrium value. In such cirumstances, a licensee is at least potentially able to appropriate, or share in, an administratively generated rent.

Such rents are competed for – by licensees, by government acting on behalf of the intended beneficiaries of subsidies, and by public officials acting personally or collectively in their regulative boards and corporations. Other economic actors attempt to circumvent official controls, and to sell or buy at free-market values. Parallel markets grow up, often unlawfully, alongside those officially sanctioned. West African governments have had particular difficulties in maintaining control over supplies of goods and the destinations of subsidies because of the imperfect policing of the long land boundaries of their states, their inability administratively to ration final users, and the tendency of official marketing agencies to become politicized where supplies cannot be allocated by price. These problems are illustrated by the official monopolization of export-crop sales, by interest-rate controls and the operation of public credit institutions, and by the distribution of subsidized imports, agricultural inputs and locally produced foodstuffs.

It may be concluded that the governments have often been too ambitious in their attempts to influence use of the resources left in private hands. They have made regulations they could not enforce, or whose consequences they could not control. Shrill official denunciations of smugglers and violators of price controls often attest to no more than the unreasonableness of official pricing of goods like foreign exchange, export crops and local foodstuffs, and to incongruity between economic policies and economic behaviour.

6

Development Policies

Introduction

All economies alter over time. They would change even if only in reaction to changes in factors external to themselves, such as annual variations in the weather, but the urge of men and women to better their material condition produces also endogenous changes. A state of economic stationariness, in which the same quantities and relationships are simply reproduced from year to year, is only a construct useful in analysing the mutations that actually occur. Whether we think of an economy as a cosmos or as a collective enterprise, it can be observed over time growing or contracting, developing or (to use an expression now fashionable) under-developing.

Development policies have two complementary connotations. Theoretically, they express distrust of privately generated economic change, and a belief that governments must intervene to ensure that changes are in the right directions, move fast enough and are not merely haphazard but mutually reinforcing. Politically, development policies denote the undertaking by governments directly or indirectly to create material welfare either for their national populations or for sections of those populations; this political commitment dates from the late 1940s in most West African countries. Thus, in an economy that is being developed (rather than developing or growing of its own accord), a government accepts or claims overall responsibility both for the management of economic change and for the delivery of material welfare. A fairly obvious consequence of this duality is that economic management becomes politicized; development policies are evaluated, and therefore shaped, by how and where and to whom they deliver welfare.

In practice, therefore, distributional questions are inseparable from development policies. In principle, this bond has not always been recognized. Economic growth has been assumed to be of common benefit, and enquiries as to distribution therefore irrelevant. Governments have been held to be capable of transcending sectional interests, and of identifying and consistently pursuing a national interest in which all personal advantages are contained.[1] In some quarters, notably the international development agencies, it seemed novel, or even revolutionary, when about 1970 attention was turned from the measurement of economic growth to the consideration of who was benefiting from the growth, and whether (for example) unemployment was being diminished, poverty alleviated and inequality reduced.[2] But in the actual execution of development policies, at any rate in West Africa, distributional issues have never been lost sight of; indeed, they have predominated in practice over the so-called growth objective. It does not follow that the policies have been implemented to diminish disparities in income or to eliminate penury; there are other distributional issues, and other conceptions of equity, than those. The predominance of distribution merely means that development policies have been regularly judged in the belief that they do not benefit everyone equally, and therefore by reference to whom they do benefit most.

On the theoretical plane, development policies attempt to accelerate the pace of positive economic change by centralizing economic decisions, or reordering the economic cosmos as a collective national enterprise. Stepping up the rate and controlling the composition of investment expenditures are one aspect of this strategy, and have been the most conspicuous feature of the development plans produced by governments. A second aspect, no less characteristic of the concept of a national economy, is the official regulation of external trade and payments, frequently with a view to substituting domestic for foreign markets, sources of supply, and factors of production. Aspirations toward national planning and national self-sufficiency, at least in what are regarded as essentials, become typical of an economy experiencing development policies.

The preceding chapter considered taxation, explicit subsidization, price-fixing and other instruments available for executing such designs, and observed particularist interests to be unavoidably penalized or advantaged by their use. Given the limitations of the instruments, development policies could not be neutral as among private interests, even if they were intended so to be. In practice, few of those responsible for shaping development policies, and perhaps none of those hoping to benefit from them, have sought neutrality in their application. Here as elsewhere, 'policy is politics'.[3]

Development objectives

In the late colonial period in West Africa, mass welfare was regarded as not so much the ultimate but the immediate outcome of development policies. In Nigeria, for example, the primary objectives of the first development plan in 1946 were better water supplies, nutrition and health. In both British and French territories, there was heavy spending on schools, colleges, clinics and hospitals and on infrastructural improvements intended to ameliorate urban living conditions, increase administrative efficiency and lower transport costs. A conscious effort was made to raise Africans nearer the norms of social provision enjoyed by residents of the metropolitan countries. In view of the enormity of the task involved, it was an effort often manifest more in the quality than in the quantity of what was provided, and there was subsequently criticism that inappropriate standards had been set in, for example, higher education, curative medicine and public buildings.

Several considerations explain the prominence given at this time to direct delivery of welfare. One is the carry-over of ideas entertained in the 1930s that the efficient functioning of the world economy required measures to raise living standards, or to satisfy basic human needs having universal validity.[4] Another is the heightened sensitivity of the colonial powers toward discharging the trusteeship they professed – a reaction to political changes in the metropolises themselves, to discontents in the dependencies, and to wartime pressures from the United States. As part of the re-interpretation of trusteeship, the doctrine that colonies ought to be financially self-sufficient was abandoned by both Britain and France after the Second World War. The British made grants to colonial administrations from Colonial Development and Welfare Funds, the French from FIDES (*Fonds d'investissement pour le développement économique et social*). That development funds could be obtained without cost helps explain the kind of uses made of them. In Nigeria and Ghana, the metropolitan grants quickly diminished in importance as export revenues grew and marketing board surpluses accrued; but those financial resources too were easily acquired, and consequently were often spent without close regard for economic returns.

The welfare bias in development expenditures was also determined by domestic politics in the period of decolonization, when representative government was being consolidated and power devolved to elected politicians. Governments in the Gold Coast and southern Nigeria spent heavily on schooling because the ability of the elected politicians to unlock opportunities through education helped legitimize their role in relation to traditional rulers and colonial officials, and because primary schools, being inexpensive to construct and relatively easy to staff, could be more widely distributed than any other amenity that was popularly demanded.[5] In AOF, Nigeria and Sierra Leone, improvements in wages and working conditions in formalized employment were conceded in recognition of the political strength of organized labour, or in the hope of securing electoral advantage.[6]

In opposition to the tenor of development policies in the 1950s was the doctrine that mass welfare depended ultimately on productiveness, and must therefore be obtained for the most part indirectly through the productive investment of the funds available for development. Though not without political attraction (as will be seen later), this doctrine had less popular appeal. It spoke of sacrifices, austerity and deferred hopes. It heightened suspicion concerning whom development would benefit, or was already benefiting. For a time, it could be warded off. Nigeria, Ghana and Sierra Leone had free reserves of foreign exchange that could be drawn on when the favourable trend in their terms of trade was checked after 1954; not until the early 1960s did finance begin seriously to constrain their pace of development. The territories of AOF, which depended more on metropolitan subventions, were careful (Guinea excepted) to maintain after independence their various aid relationships with France and later with the EEC.

Some revision of development objectives nevertheless occurred in the 1960s as a result of the increasing facility with which the administrations spent money, the wider ambitions deemed appropriate to sovereign states, and the consequential growth of scarcity in finance. This revision was most explicit in the first Nigerian national plan of 1962, and in the writings about that plan by its principal architect, W. F. Stolper.[7] Taking increase in production per head to be the supreme purpose of development planning, Stolper reasoned that decisions on the use of resources must, wherever possible, satisfy the test of economic profitability.[8] Social purposes such as the provision of wage-employment, education and better health could be more fully accomplished only by becoming more productive. Expenditures on the delivery of welfare directly could not be excluded from the plan, but they were to be regarded as non-economic – as a detraction from, or impediment to, that growth in economic production that was the indispensable condition of a better life for all Nigerians and would be achieved through the profitable investment of development funds.

Among economists, the Nigerian plan of 1962 attracted criticism, but mainly on the ground that it was not ambitious and demanding enough. Stolper's message that satisfying welfare aims was a function of production was not regarded as controversial in academic circles. The Ghanaian seven-year plan of 1963 contained a similar emphasis, holding that the long-term objectives of economic policy were to be achieved by 'maximising the rate of adoption of modern technology and the rate of productive investment'.[9] At the end of the 1960s, the permanent secretary of the Federal Ministry of Economic Development in Lagos still epitomized development as the long-run growth of output per head, regarded such growth as the foremost way in which wage-employment could be created, and warned (as Stolper had warned) that economic growth might be frustrated by premature concern with equity issues.[10]

By 1970, the identification of growth in economic production (or, more particularly, in dubious estimates of GDP per head of population) as the overriding objective of development policies, and also as the measurement of their success, was regarded as orthodoxy, at least by those who criticized these perceptions. Heresy was never extirpated. In the West African literature, a group of scholars drawn mainly from Northwestern University gave the title *Growth without Development* to their economic survey of Liberia in the early 1960s, arguing that the settler class in that country had used its monopoly of political power to reserve to itself the benefits of the rapid economic growth experienced since 1950, and had avoided policies which, through enlarging the capacities of the 'tribal' or indigenous peoples, might threaten the maintenance of its political control.[11] A little later, Samir Amin characterized the record of the Ivory Coast as growth without development, on the ground that the country's remarkable economic expansion depended on export markets and the employment of foreign capital and enterprise, and that the changes in social structure had not occurred that would permit transition to growth on the basis of domestic markets; one might speak of the development of capitalism in the Ivory Coast, but hardly of the development of Ivorian capitalism.[12] While the Northwestern team bewailed the absence of democracy, Amin deplored the lack of national economic autonomy. Subsequently, it became commonplace for commentators to distinguish between growth and development, and to claim that the former was unaccompanied by the latter wherever – rapid economic growth notwithstanding – some aim dear to the heart of the commentator was not being satisfied.[13]

A more general onslaught on the short-lived orthodoxy of the growth objective began at the end of the 1960s. The initial inspiration was the evident failure of wage-employment to keep up with the growth of production in the 'modern sectors' of the developing countries. An 'employment problem' was perceived, but surprisingly it was found not to consist primarily in unemployment. Those openly unemployed in developing countries were mostly young people, of better than average education, whose unemployment could be construed as expressing a preference for work of kinds superior to that available in agriculture or the urban 'informal sector'. Attention had therefore to be shifted from unemployment to 'underemployment'. The latter became an ascribed status. It did not denote persons working less than they wished to work but those who, however much they worked, obtained earnings deemed to be inadequate. Paradoxically, therefore, the underemployed included, and perhaps largely consisted in, persons who would normally be reckoned to be overworked. What identified them was not the amount of work they did but how little they received for doing it.

The employment problem had thus become transmuted into a poverty problem. The objection to the single-minded pursuit of economic growth was now not that it had failed to create enough wage-employment, but that it had failed to remove widespread poverty. The burden of the new, or professedly new, approach to development became the designation of poorer segments of population as 'target groups' toward alleviation of whose condition policies were to be shaped and the success of the policies judged. As a means of identifying the poor, the universal and basic human needs that had been canvassed in the 1930s were revived; the poor were those whose actual nutrition, housing, clothing, education, health care and the like fell short of the standards established, either by science or by the consensus of opinion, as necessary for health and normal human fulfilment. Satisfying these basic needs should, it was argued, be the first objective of development and the first charge on incremental production.

It became customary to speak of this approach to development serving an 'equity objective' and directed toward 'improvement' in the distribution of income. These terms were ambiguous. The view might be taken that distributive justice required equality, just as did commutative justice; but other views of economic equity (such as that people should receive what they earn) are clearly possible. Similarly, improvements in income distribution could be interpreted as the lessening of relative inequality, increases in the share of 'the lowest 40 per cent', a fall in the Gini coefficient. But reduction in inequality is not a necessary condition of poverty relief or satisfaction of basic needs[14] (nor is it likely to be a sufficient condition in countries as poor as many in West Africa, where the mean income and the mode would not be far apart). At least in principle, it is possible for poverty to be relieved, and basic needs satisfied, through absolute increases at the low end of incomes, even though relative inequality remains unchanged or actually deepens. Ought not absolute gains by the poor to be counted an 'improvement' in income distribution, even if inequality is 'worsening' (as could well be the case, if only some of the poor are gaining)? Would not an equity objective be served if the poor (or some of the poor) became better off, whether or not other groups were also gaining?

The link between reduction in inequality and poverty relief could be retained only by arguing, or assuming, that the poor could not gain absolutely except by redistribution. Failing policy changes, gains would be won only at the upper end of the income scale. Liberia as envisioned by the Northwestern team was, so to speak, the model of all developing countries; the rich got richer, and the poor were bypassed. At the same time, there was ample recognition of the practical difficulty of executing policies that would shift income and assets toward the poor and at the expense of groups who were politically powerful as well as rich. There was then a tendency to find a solution in redistributing only the increments of production. The rich would not have to sacrifice what they already had, but the additions to GNP would be distributed in accordance with the equity objective. This formula was thought, or hoped, to be politically palatable. It had the interesting implication that the discredited growth objective, far from having been displaced by the equity objective, had become indispensable to achieving the latter. Economic growth was not, after all, to be dethroned as the development criterion, but rather was to share sovereignty as William and Mary did; redistribution was to be with growth, or through growth.

The next imperative was to deny that redistribution might impede growth, and to suggest that, on the contrary, economic growth would be faster if income and wealth were less unequally distributed. It was not an easy thesis to argue. Inequality had long been regarded as economically functional in promoting saving and therefore capital formation. An even stronger case could be made for its usefulness in providing economic incentives. How, then, could growth be propelled by a reduction in inequality? One solution was to change the accounting rules; if output were valued not by its factor cost, but according to who received it, with income being weighted more heavily as one descended from richest to poorest, it would be tautologically true that reductions in inequality would increase the value of aggregate output. Though this solution made a strong appeal (and is by no means indefensible), it was not usually relied on. Use was made instead of two causal propositions. One was that redistribution of income toward the poor could be expected to produce widespread improvements in nutrition, health and education that would eventually lead to rises in the productivity of labour. The other concerned the change in the pattern of demand that would result from redistribution. The argument has been noticed in Chapter 2 of this book, although emphasis was placed there on the supposed merits of redistribution in lessening dependence on imports and foreign factors of production. The claim that the change in the pattern of demand resulting from redistribution would also induce growth follows from the observation that much productive capacity (more particularly, manufacturing capacity) in developing countries is under-utilized, supposedly for want of demand. If this deficiency in demand were corrected, unit-costs of production would be reduced, the profitability of the capacity enhanced, and the inducement to invest further in it provided. One could go further and claim that, unless purchasing power were widely dispersed, the establishment of many industries catering for mass consumption would be blocked – so long, it has to be added, as their sales were to be confined, for one reason or another, to the home market.

Thus, within a few years, thinking in the ILO and other development agencies moved from perception of an employment problem in developing countries to criticism of the growth objective, concern for alleviation of poverty, revival of the basic-needs criteria of development, support for egalitarian policies, recognition that economic growth was made more rather than less necessary by these new objectives, and strenuous efforts to show that the service of 'equity' was compatible with growth and could even lead to accelerated growth.[15]

Why did the equity objective in its various formulations leap into prominence in the early 1970s? Because of the rapid expansion of world trade since 1945, the economic growth of underdeveloped countries had not been the desperately difficult task that had been foreseen. In many of these countries (including Nigeria and the Ivory Coast), the estimated rates of growth in GDP far exceeded expectations, belied prognoses, and even defied prescriptions such as the necessity for effective central planning.

First, therefore, the international development agencies badly needed a 'new challenge' by 1970. They found it in unemployment and underemployment, the persistence of poverty, the tendency of development to bypass the masses, inequality, and unsatisfied basic needs; and a new lease of life was given their role as providers of aid, advice, research and expertise.

Second, rapid economic growth produced a more profound reaction in the form of a renewal of populist sentiment.[16] In the Third World, as earlier in the First and Second Worlds, tendencies toward large-scale economic organization and the detachment of agriculturalists and artisans from ownership of the means of production were recognized to have disequalizing effects and therefore deplored. A threatened society of independent peasants and craftsmen, using simple techniques and mainly labour of their own, was contrasted to its advantage with the emerging economy of large and impersonal enterprises, finely divided work tasks, increasing urban concentration, and nascent class struggles. This nostalgia for a more decentralized and harmonious social order required then to be economically vindicated. So, peasant farming was argued to be really more efficient than capitalist agriculture. The economic vitality of the urban 'informal sector' was discovered. 'Intermediate' or 'appropriate' technology was claimed to be more profitable than advanced technology, when factors were priced at their opportunity costs. The superior productivity of urban activities organized on a large scale was held to depend on economic policies biased in their favour. And it was argued, as already noticed, that a lower degree of inequality would make for economic growth by raising labour productivity eventually and by creating or widening markets for home industries.[17]

Such economic arguments were not merely Utopian. In West Africa, large-scale agricultural projects and settlement schemes did have a dismal record, the more so, no doubt, because they were invariably organized by government agencies.[18] Establishments in the 'modern sector' were buttressed by protection and subsidies, in spite of which small and informal enterprises proliferated.[19] Factor-price distortions probably were of some significance, though perhaps more in determining which new activities appeared profitable than in affecting technological choices in any given activity.[20]

Populist ideals also made some appeal to African governments – partly because they had also made some appeal to colonial administrations, and many official attitudes survived political independence. For example, in British colonial Africa, traffic in land had been discouraged in the interest of maintaining rights of access to this resource on a communal basis, urban in-migration had been regarded as socially contaminating, dependence on the market for food supplies had been something to be minimized, and generally the administrations had sought to control the commercialization of economic life and to restrain the emergence of economic inequalities and tension. Such attitudes have lingered in the postcolonial period.[21]

Even so, in West Africa as elsewhere in the Third World, the enthusiasm of the international agencies for equity objectives was not shared by governments. Lip-service might be paid to those objectives, but real shifts in policy were not much in evidence. The poor were not a constituency to be courted or a pressure-group to be placated; little political return was to be expected from serving 'equity' in the senses in which the international agencies understood the term. Egalitarianism was not a popular cause; not many people confused a desire to be better off with policies that would make them as badly off as everyone else. There were also practical difficulties in delimiting the poor, in countries where most people were badly off, and in finding instruments of policy that could bring them relief.

Thus, in Nigeria during the period of military government in the 1970s, there was frequent acknowledgement of the need for an incomes policy to effect 'improvements' in the distribution of income. In practice, only two forms of money-income lay within the power of the government to regulate: wages in formalized employment, and the earnings of export-crop farmers so far as they were determined by the producer prices paid by the marketing boards. Those therefore were the incomes that were raised, the former in 1970–1 and again in 1975, following the reports of wages commissions, and the latter mainly in 1974–5 in the wake of a reform of the marketing board system. The official view that these adjustments improved income distribution in some sense was not indefensible, but almost certainly there were poorer groups in Nigerian society, and in fact those being favoured had often been represented as well off by Nigerian standards.[22] An alternative way of 'correcting' income distribution, through free or subsidized public services and facilities, faced similar practical difficulties. While some of these benefits were specific as to place, making them specific as to income-group was technically (as well as politically) difficult. Improvements in communications, schooling, health care, irrigation and the supply of agricultural inputs tended to help best those who could best help themselves.

Lack of political interest in the equity objectives of the international agencies, and lack of administrative ability to execute them, did not imply the disappearance of distributional issues from development policies in West Africa. As was said earlier, questions of who is benefiting have been inseparable from the 'development process' since its inception, and there are other conceptions of equity than the alleviation of poverty or reduction in relative inequality. On the one hand, governments used their programmes of development to repay political debts, undermine political opposition, and generally to maintain political control. On the other hand, prospective beneficiaries of those programmes competed for shares in the welfare that government had undertaken to deliver. The composition of West African national societies was such that these particularist interests were commonly those of a community, clan or 'ethnic group'.[23] Those, accordingly, were the kind of interests patronized (or penalized) by development policy. The principal departure from this pattern has been a tendency to favour urban residents, or rather the residents of the capital and other large cities, since it was in those places that governments felt most vulnerable to political unrest; such preferment has been given through infrastructural spending, through subsidizing market supplies of food, and through adjustments of wages in formalized employment at times of unusual stress.

Sometimes it has been hoped that development could be detached from politics by shifting resources away from direct delivery of welfare and toward productive investment. These hopes have been illusory. Competition has perhaps been more intense for control over the means of production than for acquiring public services and amenities, for it has been rivalry among the better educated and more politically aware sections of society, a struggle sharpened by personal ambitions as well as sustained by communal interests. Governments for their part can attempt political control as well by the siting of productive projects as by the distribution of social overhead expenditures. Whatever the matter at stake – an industrial licence or a school building, an irrigation project or a clinic – what counts is having one's own man in a position of authority, whether in political office, the party hierarchy, the public service, or a parastatal body. The dominant purpose of electoral activity in West Africa has therefore been the control of such preferment. The disappearance of party politics makes little if any difference. Indeed, one-party states have resulted not only from the intolerance of governments for opposition, but also from the disinclination of the oppositions to be automatically excluded from government patronage.

The politicization of economic life through acceptance of the development ethos had become apparent in southern Nigeria and the Gold Coast by the early 1950s. Thirty years later, this feature of West African economic life shows no sign of erosion.

Development strategy: the investment fetish

The debate over development objectives conducted since 1970 has scarcely disturbed the proposition that increase in material welfare requires a greater flow of economic goods and services. The question of how this greater flow can be obtained has always been central to development theorizing and prescription. In practice, the initial solutions were provided in West Africa by Colonial Development and Welfare and FIDES grants, and by improving terms of trade. The first solution soon appeared inadequate, at least in the larger countries. The second, though it became of great practical importance, as was shown in Chapter 4, was unexpected, inconsistent with received opinion,[24] and thought to be very temporary.

The obvious candidate for the role of a more powerful, general and reliable motor of economic growth was investment or capital formation – the creation of tangible assets which, directly or indirectly, would enhance the productiveness of labour and land. Put another way, it was the inadequacy of investment that chiefly explained the poverty and backwardness of areas like West Africa, it was capital shortage that constrained their growth.

Investment was the obvious candidate for at least three reasons. First, a belief was strongly entertained among economists in the late 1930s and early 1940s that investment opportunities had become chronically deficient in the rich, industrialized countries of the Western world, such were the wealth and the saving propensities of those countries. The counterpart of this view was that in other, much poorer, parts of the world savings must be low in relation to investment opportunities. This contrast presented interesting possibilities of capital transfers that might be profoundly beneficial to both parties, the rich being disembarrassed of their superfluity, the poor relieved of their want. 'World development programmes' to accomplish such transfers were accordingly canvassed during and immediately after the Second World War.[25]

The second reason, like the first, was related to the evolution of macroeconomic thought. Work had begun in the 1940s on the 'dynamizing' of Keynesian theory so as to take account of the effect of increases in investment spending not only on effective demand (as in Keynes's short-run model of income determination) but also, in the longer term, on aggregate supply. The growth models thus produced, and associated with the names of Harrod and Domar,[26] postulated capital-output ratios in the sense of determinate, numerical relationships between additions to a stock of capital and ensuing increases in a GDP. In its pristine form, the incremental capital-output ratio was represented neither as an economic law nor as relevant to economically underdeveloped countries. But planners of economic growth in those countries found the concept irresistible. In the hands of its most naive users, it was capable of reducing all the questions of how to engineer growth into a single, simple imperative: raise the investment rate. It was also the means of calculating the 'investment requirement' of a target rate of growth in GDP. Assuming that projected domestic savings fell short of this requirement, it was the means of quantifying a country's need of external aid.

International aid was the third reason why capital formation was recognized as the propellant of economic growth in underdeveloped countries. Once the case for aiding the development of such countries was accepted by the richer governments (as in the UK in 1940, France in 1946 and the United States in 1949), the ways in which aid could most readily be given helped shape the understanding of why poor countries were poor. The rich could supply producers' goods as grants or on concessional loan terms, and the men and women whose technical knowledge, skills and organizing ability would allow absorption of those assets by the economies of the aided countries. Accordingly it was the lack of capital in those senses that explained economic backwardness. Any other explanation tended to be excluded by the inability of capital transfers to remedy it. Investment was the key to growth because investment was what aid could increase. A sufficient volume of investment would eventually make aid redundant, through imparting such momentum to an economy that it would 'take off' into self-sustained growth.[27]

These three factors were mutually reinforcing in establishing investment as the propelling force of economic growth in areas like West Africa – the supposed superior productivity of capital in lands where capital assets were few, the making of causal and quantifiable connections between net investment and increases in aggregate output, and the availability of foreign aid to supplement domestic savings in financing investment. Intellectual and professional interests (to say nothing of business interests) became entrenched in this perception of the growth process. 'A generation of planners and foreign aid officials came to believe in the reality and the manipulability of the Propensity to Save and the Capital-Output Ratio, and they stuck to this faith over an astonishingly long period of time for the good reason that the representation of the world in terms of these concepts was essential to their status as experts – it was "the only game in town".'[28] In 1971, it could still be stated accurately that: 'The dominant and orthodox economic theory emphasises capital formation as the main factor in development of the developing countries. This theory is taught to most students and is the basis of most comprehensive economic development plans prepared by developing countries.'[29] Though less is heard nowadays of predictable numerical relationships between net investment and economic growth rates, it remains true that the investment rate is generally regarded as profoundly significant in determining the growth prospects of any country.

Like several other tenets of development theory, this principle is supported in disregard of historical evidence. For the data available for the industrialized countries since the late nineteenth century indicate that the mere growth of capital (and also of the labour force) contributed little to the rise in output per head. Kuznets, for example, drew from this historical evidence 'the inescapable conclusion . . . that the direct contribution of man-hours and capital accumulation would hardly account for more than a tenth of the rate of growth in per capita product – and probably less.'[30] Some 'residual factor', raising the quality or efficiency of the accumulating productive factors, had evidently been vastly more important as a source of economic growth than had quantitative change in the stock of capital and the labour force.

These findings might have been expected greatly to weaken the standing of capital formation in the growth process. In fact, they led instead to a further extension of the meaning of investment.[31] The residual factor was postulated to consist mainly in the growth and transmission of knowledge, becoming embodied in tangible assets and human 'inputs' of continually improving quality and growing productiveness (educationists, in particular, favoured this interpretation). Though it would be misleading to suggest that knowledge had previously been neglected – technical assistance, for example, had figured conspicuously in both the advocacy and the practice of international aid – expenditure on developing human capacities acquired new prominence about 1960 as a condition, instrument or determinant of economic growth; 'human capital' was rediscovered.

Since human capital is a highly elastic concept,[32] a great variety of expenditures might now have been thought eligible for inclusion in the 'capital requirement' of a target rate of economic growth. In practice, admission was usually limited to spending on advanced education. The landmark in West Africa is the report, published in 1960, of the Ashby Commission on Higher Education in Nigeria. 'In order to achieve a rate of growth in the next decade equivalent to that of the past,' Frederick Harbison advised the Commission, 'certain "inputs" of capital and high-level manpower are required. The targets set forth in this report represent the *probable minimum "inputs" of high level manpower* required to permit a rate of economic growth in the next decade equivalent to that in the ten years before independence . . . If the economy is unable to achieve the degree of high-level human resource accumulation as represented in general by these targets, then it may have to settle for a more modest rate of economic development. Likewise, capital cannot be productively employed in Nigeria to promote economic growth unless at the same time the required high-level manpower is forthcoming.'[33] A connection was thus established as confidently between one kind of human capital formation and the growth of economic output as it had earlier been established between physical capital formation and the growth of output. When Harbison wrote later that 'the rate of net accumulation of high-level manpower for expansion activities is calculated at 1.5 times the increase in GNP,' and that 'in most cases, the number of people in the intermediate category should increase twice as fast as GNP,'[34] it was made explicit that he was using incremental capital-output ratios, with the prefix 'human' now attached to the capital.

Thus, economic growth was held generally to depend on, or to be impossible without, increase in the stocks both of material assets and of formally educated manpower, and these functional relationships were further believed to be quantifiable. Additionally, it was held that investments, whether in material or human capital, required to be balanced in relation to the composition of future output. The latter would depend on relative income-elasticities of demand for final products, on input-output coefficients (including import coefficients) relating to intermediate products, and on export projections. If the target output were thus determined not only in total but also in composition, the 'investment requirement' could be similarly disaggregated on the basis of the incremental capital-output ratios for particular products or industries. Obviously, investments would then have to be centrally coordinated or planned as they were undertaken.

These were not the ways in which investment opportunities had been discovered in the past in West Africa, or perhaps anywhere else in the world, and their adoption would clearly make enormous demands on the ability to collect and process information and to enforce the conclusions reached by the planners. Nevertheless, it was believed that centralization of investment decisions was essential in order to reveal their full social profitability (for the profitability of any one investment would usually depend on whether certain other investments were being undertaken) and to ensure that the investment rate was as great as was socially desirable. Without a plan, confidence in the growth of markets would be lacking, outputs would fail to dovetail with inputs, and the inducement to invest would be deficient. Such, in desperate brevity, was the theoretical case for comprehensive investment planning that was argued in the 1940s and 1950s and carried much conviction.[35]

Planning of this nature was always a pretence in West Africa. Plans began to appear immediately after the Second World War, but they were little more than edited compilations of the desiderata of government departments, produced as a condition of sharing in metropolitan aid funds – colonial shopping lists, as they were later dismissively termed.[36] While the staking of claims to external finance remained an important latent function of planning, later plans were made with much more technical sophistication; their apogee was perhaps the Ghanaian seven-year plan of 1963, which had the distinction of endorsement by a conference of economists brought to Accra from many parts of the world. These professedly comprehensive plans nevertheless excluded large areas of economic life, notably the production of and trade in foodstuffs, implicitly on the ground that they neither experienced nor required investment, and private economic enterprise in general received little more than token recognition.

Further, whatever the macroeconomic merits of these plans, they were characteristically deficient in the preparation of the projects by which investments in the sense of monetary allocations were to result in additional flows of goods and services. Much planning proceeded 'without facts' (to use Stolper's phrase),[37] or in ignorance of how the sums notionally available to particular activities could be used in practice to raise their outputs. The greater the internal consistency of a plan, the less its operational content tended to become. In this respect, the plans which emphasized coordination of industries and sectors compared unfavourably with the disparaged colonial shopping lists, for the latter had been essentially collections of projects.

Finally, the discipline required if central planning was to be effective was lacking in West Africa. The reason was not the resistance of private enterprises to this attempt to superimpose a higher rationality on their own, for the private sector remained largely outside the plan (though its behaviour might be powerfully constrained by attempts to implement the plan). Indiscipline resulted rather from the refusal of politicians and administrators to tolerate, through adherence to the plan, the loss of their autonomy, the removal of their discretion, the disappearance of their options.[38] It may be blamed on a refractory political culture, or (more reasonably) on the folly of planners in abstracting from politics when designing programmes that could be executed only by political will.[39]

At best, therefore, central plans in West Africa have been loose envelopes of ambitions and desires within which the economies have evolved; at worst, they have been totally irrelevant to what actually happened. A survey of planning in Africa published in 1965 by two staff members of the UN Economic Commission for Africa observed that 'the hold of the plan on the nation's economy is often almost negligible,' and that, 'if any of the projects which are incorporated in the plan are fulfilled, this is often fortuitous.'[40] In a wider context, it was said of national planning in 1969 that 'nobody working in this field can fail to be struck by the contrast between this intellectual input and the actual results,' and that some plans had been 'little better than fantasies'.[41] More specifically, it has been said of the celebrated Ghanaian seven-year plan that this was 'a piece of paper, with an operational impact close to zero', and that 'we see an almost total gap between the theoretical advantages of planning and the record'.[42] National plans continue to be produced, but they are significant only of the aims and expectations officially entertained when the plans were being drafted; 'I know of no African state which is currently engaged in a serious planning effort,' wrote one authority in 1976, 'in the sense of using its plan as a guide to day-to-day policy decisions and the preparation of its budgets.'[43]

While the theory of comprehensive planning, showing the necessity to centralize investment decisions, has had little or no effect on the allocation of investible resources in practice, it has served a purpose in helping rationalize and legitimize the precept to increase investment. The amount of investment and the size of the plan became interchangeable ideas, imparting to the former the allegedly scientific properties of the latter. The postulated causal connection between investment and economic growth was grasped with such fervour that diverting resources from the satisfaction of current wants was regarded in some quarters as a social benefit rather than a sacrifice, and consumption a waste rather than 'the sole end and purpose of all production' (as Adam Smith described it). This attitude was the result of excluding, consciously or unconsciously, any alternative sources of economic improvement.

Thus, Gunnar Myrdal, for whom central economic planning was 'the first condition of progress', believed also that there was 'no other road to economic development than a compulsory rise in the share of the national income which is withheld from consumption and devoted to investment'.[44] Such beliefs, which make of investment or the plan a kind of fetish demanding regular and increasing propitiation, are by no means obsolete.[45]

It is notable that increases in capital formation were expected to require compulsion, and not to be achieved by the attraction of prospective rates of return on investments. Savings would be enforced by surpluses of budgetary revenue over current expenditure, by the profits of state monopolies, or, failing these resources, by the 'inflation tax'. Compulsory funding also describes the investments financed by aid, for aid was transfers which taxpayers in the donor countries were obliged by their governments to make.

It is notable too that the proceeds both of enforced savings and of aid would be at public disposal. Hence the economy would become more collectivized as it became more capitalized. The consequences for private economic motivation and behaviour were thought unimportant. 'It is shortage of resources, and not inadequate incentives, which limits the pace of economic development,' wrote Kaldor. 'Indeed the importance of public revenue from the point of view of accelerated economic development could hardly be exaggerated.'[46] Helleiner's justification of the marketing boards conceded that the immediate effect of the heavy taxation of export-crop farmers might have been to reduce export earnings, and therefore GDP, but argued that this was a price worth paying for the sake of a 'superior' distribution of export earnings, public agencies having intrinsically better uses for funds than the farmers had.[47] As these illustrations suggest, raising the investment rate and enlarging public command over resources were regarded as much the same thing.

Relatively to richer countries, those of West Africa could reasonably be said to be under-capitalized and under-governed in the 1950s when development policies were forged. Richer countries did have higher ratios of investment and of government spending to GDP. More particular positive correlations could be found between levels of GDP and relative measurements of (for example) manufacturing output, university enrolments and electricity generation. Development theorists and advisers were too ready to interpret these correlations as unicausal connections: to suppose that increasing the proportion of investment, government spending, manufacturing output or whatever was certainly a necessary, and possibly a sufficient, condition of greater GDP. The fallacy was not unrecognized. Attempts were made to guard against it by stipulating that investments must be productive, plans internally consistent, and public resources wisely channelled. Such prescriptions became tautologous; they were equivalent to saying that GDP would be raised by expenditures that raised GDP. They did little in practice to limit capital spending to investments that really did enhance the productiveness of labour and land, or to restrict government outlays to those that were wise.

It therefore became possible for budgets to be expanded and investment rates raised with little positive impact on the rate of economic growth, and sometimes even with negative consequences. This feature of the so-called development effort became apparent at an early stage in West Africa. It underlay the apprehension that inappropriately high standards were being set in infrastructural provision. In Nigeria, the coincidence was perceived of a rising investment rate and a declining rate of growth of output in the 1950s, and the fear expressed 'that a constant stream of investments would simply lead to a constant level of income' unless something was done to alter the composition of investments.[48] Ghana in the 1960s provided a more dramatic case of diminishing returns from capital formation. Between 1960 and 1965 gross fixed investment averaged 18 per cent of the GDP estimate, and the capital stock is estimated to have been increased by 80 per cent (having already been enlarged by 50 per cent in the second half of the 1950s), yet the estimate of real GDP (adjusted for terms of trade effects) grew so slowly that 'attempts to measure incremental capital: output ratios for this period result in figures which are either absurdly large or even negative, and . . . even the *average* ratio doubled between 1960 and the peak year of 1967.'[49] 'The overwhelming impression is of failure – to generate growth, to invest resources wisely, to raise productivity . . . High levels of capital formation failed to generate growth either in the short run or later in the decade.'[50]

The 'paradox of investment without growth'[51] is puzzling if growth models are believed to approximate reality, central planning to be efficacious, and governments to be single-minded in their pursuit of a larger GDP. Otherwise, there is nothing astonishing in the discoveries that mistakes are made in creating capital, that costs are sometimes underestimated and prospective returns exaggerated, that some funds are sunk with little or no regard for economic profitability, and that the possibilities of waste and error are compounded as governments assert economic control and investment becomes politicized. In these respects Ghana in the 1960s is not a curiosity. Much investment in, for example, the Ivory Coast, Nigeria, Sierra Leone and Togo has been no less inconsistent with the expectations of the economic model-builders.

These false expectations can be traced to a simple theoretical misconception. When economic growth was held to depend on the rate at which resources were transferred from unproductive or less productive uses ('consumption', or the private sector) to more productive uses ('investment', or the public sector), it was implied that these better uses were known and their adoption restrained only by insufficient command over resources. This was a classical conception of the growth process, refurbished in the context of developing countries by Arthur Lewis,[52] making growth dependent on the 'capitalist surplus' currently forthcoming. With it may be contrasted the view that the principal determinant of growth is the economic exploitation of newly discovered uses for resources, the diverting of resources to these uses being assumed to occur without prevision merely through the pull of profitability. According to this view, growth depends on developing new products, factor combinations and markets, rather than on accumulating resources for making established products by familiar methods for the satisfaction of existing wants.[53] The emphasis on qualitative changes is compatible with the empirical findings, mentioned above, concerning the sources of economic growth in the industrialized countries since the late nineteenth century (it can also be said to square with everyday observation). From this standpoint, the crucial constraint is not savings but innovative ability or (in a narrowly defined sense) entrepreneurship, not the supply of resources available for better uses but the demand for them.

If resources go on being mobilized for investment in spite of a constrained demand for capital, the effects on growth are bound to be disappointing and possibly immiserizing. Evidence of deficiency in the effective demand for capital in West Africa is abundant. As mentioned already, central plans have typically been lacking in feasible projects, and spending under these plans has been distorted by the difficulties of investing in directly productive activities relatively to other sectors; planning has failed in its ostensible purpose of strengthening economically the inducement to invest. Rather than create new productive assets, governments have often preferred to use investible resources in purchasing privately owned (more particularly, foreign-owned) assets. Disbursements of foreign aid have regularly fallen short of commitments for want of schemes satisfying donors' ideas of value for money, and aid agencies can be observed competing for the limited outlets they can find for their funds.[54] Official lending agencies, established to help in the finance of indigenous businesses, have had difficulty from the beginning either in finding commercially acceptable borrowers or in recovering with interest the loans they have made; the shortage of capital which these businesses have professed has consequently appeared illusory.[55] Productive capacity has been created only to be used persistently far below its potential, especially in manufacturing, or even left entirely unused;[56] ironically, even patterns of social behaviour appear to inhibit the employment of capital more severely in West Africa than in the United States where capital is said to be abundant.[57] To an uncertain degree, inflows of both financial capital and high-level manpower have been offset by outflows, as West Africans have found better uses for their money and talents abroad than at home.

Additional evidence of a deficient demand for capital consists in the acknowledged need to enforce domestic savings and to secure international transfers on concessional terms. Domestic savings and international transfers do not need to be obtained by compulsion so long as attractive investment opportunities are available. It follows that there is a tendency for collectivized savings and aid to be used in investments which cannot otherwise attract finance, being of prospectively low productivity or having exceptionally uncertain outcomes. It may be argued that governments, if they are to discharge their recognized responsibilities, are bound to create assets in the public sector which are, at best, only indirectly and uncertainly productive, and that they may find enforced savings or aid to be the only practicable ways of financing these expenditures; but this argument merely concedes the point that the connection between so-called investments and the growth of output is problematic, if not altogether absent, in important sectors of the economy.

Expectations of economic profitability have undoubtedly underlain many of the capital outlays made in West Africa in recent times, including foreign direct investments in mining and private (and often non-financial) investments in agriculture and the urban informal sector. A greater volume of capital expenditure, including most of that made by governments, has been instigated less by expectations of gains in future output than by the gains in disposable income currently being experienced; investments, that is to say, have been determined by income rather than the other way round. This sequence is familiar in private investment in housing. It explains also much of the government spending on capital purposes in West Africa. Public expenditure on economic and social infrastructure has risen, sometimes along with public participation in the capital funding of directly productive activities, mainly because of increases, particularly sudden increases, in the resources at public disposal, rather than because of renewed dedication to the objective of economic growth. This is the way in which extremely high investment rates have been attained in several West African countries (the Ivory Coast, Mauritania, Nigeria, Togo), as was shown by the estimates in Tables 5 and 6 above. Usually these windfalls have resulted from favourable movements in the income terms of trade, tapped by fiscal regimes closely connected with the value of external trade and supported by marketing boards or stabilization funds; occasionally, as in Mauritania in the mid-1970s, the cause has been relatively massive injections of foreign aid. The outstanding case has been Nigeria in the 1970s. The combination of a rising trend in the volume of oil exports, sharp rises in their unit-value and increasing taxation of the oil industry produced an expansion more than twentyfold (at current prices) in federal government revenue between 1970 and 1980. The consequence was an increase in the ratio of development or capital spending to total federal spending from less than one-fifth in 1970 to around two-thirds in the second half of the decade.

Capital formation continues to be offered as the motor of development in areas like West Africa in spite of powerful objections – the practical irrelevance of capital-output ratios and of the methodology of central planning in general; the fallacy in deducing high productivity of capital from sparsity of capital assets; the inconsistency with that deduction of the acknowledged needs to enforce savings and to obtain external aid; and the tendency for investment to move as a dependent variable of income rather than vice versa. The doctrine that investment governs growth is both unhistorical, as the empirical studies of industrialized countries have shown, and, from an economic standpoint, untheoretical, since factor proportions (combinations of material assets and highly educated manpower with other labour and with land) are being postulated without reference to relative factor costs. On both empirical and theoretical grounds, capital formation can more reasonably be regarded as a follower than as a leader in the process of economic growth.

The doctrine survives not only because of intellectual conservatism. Whatever they do or fail to do to the rate of growth, government investment programmes certainly provide business for consultants, contractors and suppliers, both at home and overseas, and they put power into the hands of politicians and administrators. Ambitious plans are therefore not in want of political support, nor do they lack attraction as means of political control. As was suggested earlier, supposedly productive investments can be politicized as thoroughly as direct deliveries of welfare – perhaps more so, since the prospective beneficiaries belong to the most articulate, politically organized and influential sections of society. In West Africa, government contracts have become central among the prizes sought through political competition.[58] Particularist interests, both domestic and foreign, are entrenched in, and therefore help sustain, high rates of public investment, the enforcement of savings and the flow of international aid.

Development strategy: collective self-sufficiency

Among individuals, self-sufficiency is a recipe for poverty, the fate of a Robinson Crusoe. Its consequences for small communities are little better; the poorest peoples in West Africa are to be found among those who are economically most isolated. Yet collective self-sufficiency is among the leading elements of development strategy in Africa – admittedly, not usually in an absolute sense (though that is what proposals to 'de-link' from the world economy suggest), but at least to the extent of making external trading connections and the use of foreign factors of production subject to discriminatory regulation by national governments.

These policies directed toward economic autonomy are a reaction from the colonial order, and they could as well be attributed to desires to establish national identity as to objectives of economic growth. Nevertheless, the policies are widely believed to be justifiable by their effects on growth and even, in some circles, to be (like a high investment rate and central planning) an indispensable condition of growth.

Thus, according to the neo-Marxist theories of economic imperialism, Africa was fashioned as a destination for metropolitan savings and products and a source of cheap raw materials and foodstuffs in order to retard declining profitability or avert commercial crises in Europe. It became an appendage to a system of economy centred elsewhere, adjusting its occupations and distribution of population, its tastes and ways of life, to suit the convenience of outside capitalist interests. Far from assisting African development, this relationship of dependence is held to have set in motion a process of underdevelopment. Political independence provided the opportunity to escape this subordination. Where this opportunity was not taken, the failure was evidence that the process of underdevelopment continued, outright colonialism having been succeeded by disguised, or neo, colonialism.[59]

There are obvious objections to this interpretation. For Africa to be used to alleviate the 'contradictions' of capitalism, it needed to be of quantitative importance in the world economy, but in practice its contribution to international trade, and its significance as a destination for capital flows and migration from outside Africa, have never been great. Again, some African economies undoubtedly expanded greatly in the colonial period, the process of underdevelopment notwithstanding, and some have gone on expanding rapidly in the neo-colonial period.[60] So, granting their dependence or subordination, this relationship has not necessarily been disadvantageous unless 'development' is defined to imply territorial autonomy (in which case the argument becomes tautologous), or unless the counterfactual (and therefore unverifiable) assumption is made that the expansion would have been even greater in the absence of colonialism. With reference to the latter point, one might observe that the neo-Marxist interpretation scarcely deserves its name, since Marx himself believed colonialism to be a constructive, liberating and 'progressive' force.[61]

The belief is nevertheless profoundly influential that the unregulated incorporation of African countries into the world economy has prejudiced their economic growth. This incorporation caused resources to be deflected into primary production for export, and therefore (it is argued) out of manufacturing, the more sophisticated kinds of services, and even food production for local consumption. This pattern of resource use need not be attributed to imperial policy. It occurred merely as a reaction to current comparative advantage – the export of beverage crops, raw materials and minerals being the most remunerative available uses of labour and natural resources. Similarly, although the 'polemics of imperialism' can be cited as evidence of official discouragement of large-scale manufacturing in colonial territories, the absence or rarity of these activities can be readily explained by comparative advantage; most manufactures could be more cheaply obtained from abroad than produced locally.[62] Food imports, and reliance on foreign services such as shipping, banking and insurance, are no less amenable to economic explanation. But the consequences of thus exercising economic rationality were economies that appeared unbalanced, lopsided or structurally distorted, and whose malformation was held to reduce their performance below what might have been attained and strongly to inhibit their further growth.[63]

The unbalancing of African economic structures through comparative advantage is held to have had negative consequences for three principal reasons. First, so long as external capital, enterprise and labour were needed to create and maintain the export industries, either directly (as in mining, timber extraction and plantation agriculture) or indirectly (as in the export and import trade and financial services), economic surpluses tended to be drained out of the African territories and investment in them was therefore checked. Second, export trade in primary products was seen as peculiarly disadvantageous because the prices of those products were abnormally unstable, making an orderly planning of economic growth impossible; alternatively or additionally, those prices were undergoing secular deterioration relatively to the prices of the imports of manufactures received in return, so that equal volumes of exports bought less and less as time went by. Third, the absence of a substantial manufacturing sector was regarded as a handicap, partly because the prices of manufactures must be improving if those of primary products were deteriorating, but more importantly because large-scale manufacturing was regarded as a modernizing influence *par excellence*; it was 'a central factor of dynamic radiation' in economic life, which, more than any other activity, provided the growing points for technical knowledge, education and the development of skills.[64]

Shortfalls of national disposable from domestic income, attributable to the use of foreign-owned factors of production, were discussed in Chapter 2, and were seen to be substantial in a few

West African countries including the Ivory Coast and Liberia. As was pointed out then, such 'losses' may be compatible with levels of national disposable income, of national savings, and of foreign exchange earnings net of factor payments and transfers abroad, which are higher than could otherwise be attained. It is not self-evident that national could be substituted for foreign factors of production without prejudice to short-run efficiency or growth prospects or both. Nor is it clear that checking the 'drain of surplus' through a fundamental restructuring of the economy could be achieved without such sacrifices. In any event, the role of the surplus, or of investible funds, in propelling growth is much less than it is often represented to be, as has been argued earlier in this chapter. Even more obviously fallacious is the popular argument that the negative effects of foreign participation are proven when the outflow of interest, dividend and amortization payments on foreign capital exceeds the inflow of new foreign capital, for this comparison tells nothing about what effects the foreign participation has had on productivity and incomes in the host economy, although these effects are precisely the matter at issue.

Discussion earlier in the present chapter will have suggested that there are many more obstacles to the orderly planning of growth than fluctuations in export earnings. In Chapter 2, it was pointed out that these fluctuations, attributable to instability in export prices or to other causes, can be compensated through stabilization funds or by commercial borrowing abroad. Is this source of economic instability nevertheless so important as to justify a government in discouraging exports, or in attempting to diversify exports by diverting resources to products in which comparative advantage is less? This reaction would seem to be rational only if the growth path of the more secure national income (or, perhaps more particularly, of the more secure government revenues) were to be not significantly below the trend in unstabilized income or revenues.

Grounds for pessimistic projection of the trend in these unstabilized quantities were supplied by the thesis of secular deterioration in the terms of trade of developing countries (or, more accurately, of primary-product exporters). This thesis was based on statistics relating to UK export and import prices between the 1870s and the Second World War. It was held to have received further empirical support in the ten or fifteen years following the break in the mid-1950s in the postwar boom in commodity prices.

There is no doubt that its political acceptability in developing countries explains most of the support attracted by the thesis. Properly interpreted, the statistical evidence provides at best uncertain validation for the period preceding the Second World War, and no validation at all for the postwar period.[65]

Changes in the ratio of export to import prices (the commodity, or net barter, terms of trade) are in any case a misleading indication of the gains won from international trade. The commodity terms might deteriorate while the single-factoral terms of trade (allowing for changes in export productivity) or the income terms of trade (allowing for changes in export volume) were improving. The cheapening of exports could then be compatible with increase in the profitability of export production, the aggregate of export earnings, or both. Data on export productivity are scarce for West Africa, but it is generally believed not to have been buoyant. On the other hand, Table 16 and Figures 1 and 2 above do show several instances where estimates of the commodity and income terms of trade of particular West African countries have moved in opposite directions, and other instances where both the commodity and the income terms have improved but the latter much more than the former. Oscillations in the commodity terms notwithstanding, it is clear that the aggregate purchasing power of West African exports over imports has been vastly enlarged in the period since the Second World War. This expansion is the more remarkable since policies of self-reliance have been frequently defended on the ground of exceptional difficulty in earning foreign exchange, and since exports have in fact been discouraged by discriminatory taxation and, in some countries, by overvalued currencies.

There remains the argument that a country is economically penalized by the absence or underdevelopment of a large-scale manufacturing sector; thus denied, it will remain technically unprogressive, antiquated in its culture, backward and poor. The belief that foreign trade is responsible for this handicapping implies that a path of economic evolution centring on manufacturing for home markets would otherwise have been followed. This implication is questionable; in West Africa, indeed, it is extremely implausible. Mass-produced manufactures require markets, which are provided by rising income. Income has been raised in West Africa mainly by exports of primary products, from palm oil to crude petroleum. So far as creation of demand is concerned, the export trade has been the path to industrialization rather than a diversion from it; thus, foreign investment in manufacturing in Nigeria and the Ivory Coast has been chiefly motivated by the growth of the market.[66] For manufacturing itself to have been no less effective in creating the market for manufactures, it would have had to produce its goods as efficiently as they could be obtained as imports through the production of primary-product exports. In the first half of the twentieth century, very few people in West Africa believed satisfaction of this requirement to be possible, which is why it was so rarely attempted. Counterfactualism is pushed to the extreme by the belief that the path of growth through home-market manufacturing was available in West Africa as an alternative to growth through exports.

Government measures to hasten industrialization in disregard of economic efficiency can nevertheless be defended as potentially fruitful. Because of a protective tariff, and possibly explicit subsidization, an industry is established sooner than would be warranted by strictly commercial considerations. The financial losses occasioned by this premature action may then be more than offset by the external economies which the industry proceeds to generate – upgrading labour skills, disseminating technical knowledge, inducing investment opportunities in related activities, and so on. Possibly, also, the inefficiency of the industry relatively to competing imports will be temporary; it may disappear as experience is gained, economies of scale are won (assuming the market expands further), and infancy is outgrown. According to this view, comparative advantage is certainly not irrelevant to the possibilities of economic growth, but it needs to be interpreted 'dynamically', recognizing that countries can reshape their comparative advantage over time by the investment decisions they make (on this ground, as was seen in Chapter 4, customs unions of African countries have been defended). It might be puerile to suggest that Ghana or Senegal could have chosen to become manufacturing nations fifty or a hundred years ago, but this is not to say that their industrialization could not be accelerated by policy measures, once they were freed from colonial control.

Despite its long pedigree, the infant-industry argument provides a poor case for forcing the pace of industrialization. If falling costs can really be anticipated, initial losses will be compensated by future profits, and there is no reason except special pleading why investors should be relieved of those losses. If the fall in costs cannot be confidently foreseen, the infant has to be regarded as a long-term or permanent dependant, and the case for supporting it reduces to the externalities with which it is credited.

Externalities may be important, but they are inconstant, difficult to measure, not always positive, and not peculiar to manufacturing activity. Further, their fair compensation would require a more exact and flexible fiscal administration than those possessed by West African countries, and perhaps any other countries. In principle, a system of subsidies and taxes can be envisaged that would adjust (and readjust) the private profitability of any activity in line with its full social advantage. In practice, devices like protective tariffs, other restraints on imports, government loans, input subsidies and tax remissions are inexact and often undiscriminating means of delivering these compensations. They are also the outcomes of political processes, not detached analysis. These adjustments accordingly owe more to bargaining, special pleading and political patronage than they do to economic theory and measurement – though the economics plays, as is so often the case, a useful rationalizing role.

The case for forced industrialization – or, more generally, for policy-contrived diversification of economic activity – is a case for limited and highly selective departures from current comparative advantage. The limits are imposed by the costs of diverting resources from their immediately most remunerative uses; the selectivity is required because there are probably few new activities whose social profitability is substantially greater than their private profitability, and because the means of correcting such divergences are in any case crude. An unrestrained and unselective forgoing of comparative advantage does not promote economic growth, not even in the long term. Unless they confer genuine and substantial external economies, inefficient import-substituting industries created at the expense of other activities are liabilities, not assets. They are pools of privilege, not springs of progress. They may serve non-economic purposes, like enhancing national sovereignty or strengthening political control, but they will not produce economic growth.

Diversification has often been contrived in West Africa with scant regard for economic gain or loss, or on the basis of unrealistically high evaluations of externalities. One measurement of the results is given by the domestic resource cost (DRC) of saving (or, for exports, earning) foreign exchange. It is the ratio of the opportunity costs of all the domestic resources used, directly or indirectly, in producing a unit of a good to the net foreign exchange so gained. The latter quantity is the difference between the gross foreign exchange savings (or earnings) and the foreign exchange costs directly or indirectly incurred in the production.

Ghana in the early 1960s is a celebrated case of accelerated industrialization; its president took the view that its backwardness resulted from being 'the dumping ground of other countries' manufactures', and that 'if we were to refrain from building, say, a soap factory because we might have to raise the price of soap to the community, we should be doing a disservice to the country.'[67] W. F. Steel estimated DRCs in 1967/68 for a sample of forty Ghanaian firms, broadly representative of large-scale manufacturing, and compared them with the official exchange rate, then 1.02 cedis to the dollar, and with a shadow exchange rate of 1.53 cedis to the dollar (assuming the cedi to be overvalued by one-third). Only six of these firms (and 13.5 per cent of the total output of the sample) were efficient in the sense that they were transforming domestic resources into foreign exchange at a rate equal to or less than the value officially put on foreign exchange. A further four firms (and 19.2 per cent of the output) became efficient at the shadow exchange rate representing the possible equilibrium value of the cedi. The other thirty firms were inefficient in the sense that it cost them more to save foreign exchange than foreign exchange was worth even at the shadow rate. They accounted for 67.3 per cent of the output, and they included eight firms (24 per cent of the output) which would have been inefficient at *any* rate of exchange since their net foreign exchange savings were negative.[68] Steel found no evidence of externalities sufficient to justify the high DRCs. Dis-saving of foreign exchange was not confined to some import-substituting manufactures; it appears to have been achieved also by the processing of export commodities and by the modern fishing fleet,[69] and probably by the air and shipping lines.

In the Ivory Coast, industrialization has been less strenuously pursued, and the protection given domestic manufacturing is much less marked. Even so, a study by the World Bank in 1975 of eighty-nine modern manufacturing firms produced an overall DRC coefficient (ratio of the DRC to the official exchange rate) of 1.34 – i.e., it cost about one-third more to gain foreign exchange by domestic manufacturing than foreign exchange was worth at the official rate (and in this case the official rate was probably not greatly below the equilibrium rate). It may be thought that such a margin of extra cost might be justified by externalities. It should therefore be noted that the DRC coefficient varied widely among industries; it appeared well below unity in beer, soft drinks and board and paper articles, but was as high as 3.33 in flour milling, 3.16 in footwear and 2.31 in textiles and clothing.[70] External economy effects there might be, but it is unlikely that they were so variously distributed among industries as these figures would suggest.

Valuing domestic resources at market prices instead of opportunity costs, the DRC coefficient becomes a coefficient of effective protection (EPC) – the proportion by which value added in a domestic activity is allowed by protection against imports and by domestic price fixing to exceed this value-added measured at 'border' prices (or import costs). The EPCs in Nigerian manufacturing were estimated in 1968 to be 2.2 for textiles, 2.43 for metal goods, and over 3 for furniture, glass products and radio and television assembly.[71] These rates will since have fluctuated with changes in the Nigerian import regime attributable to swings in the balance of payments, but a tendency toward increase in the EPCs, and the associated DRC coefficients, must have been created by the more recent promotion of intermediate manufactures (steel, petrochemicals, and semi-products like vehicle components). Inefficiency is likely to be initially even greater in production of these goods, because of the greater significance of economies of scale, than in the later stages of manufacturing. Attempts to accelerate integration of the manufacturing sector are an understandable reaction to the discovery that later-stage manufacturing operations contribute little toward self-reliance, being still heavily import-dependent. But this further layer of premature industrialization must, once again, be seen as a drag on economic growth, unless an extremely sanguine view is taken of the externalities it will generate, or appropriate assumptions are made about future trends in the terms of trade.

Industrialization policies have lost much of their standing in the international agencies as a result of exposure of the high costs involved in departing from comparative advantage,[72] and also because of what was termed, earlier in this chapter, a populist reaction from concentration of the forces of production and from urbanization. National governments have been less easily persuaded of their errors. Indeed, African governments have been strong in their advocacy of self-reliance in yet another dimension, that of food production, since the early 1970s; it was then that

'Operation Feed Yourself' was launched by the military government in Ghana, followed by the National Accelerated Food Production Programme and 'Operation Feed the Nation' in Nigeria. Self-sufficiency in at least basic foodstuffs is now declared to be an objective by nearly all African governments, and is one of the very few specific targets of the Lagos Plan of Action adopted by the members of the Organization of African Unity in 1980. This aim is often said to be dictated by evidence of nutritional deterioration in Africa. Apart from the objection that this evidence is slender, it is inconsistent with economic reasoning to suggest that such a decline can best be arrested by using resources directly to satisfy nutritional wants.

Food imports into West Africa grew rapidly after the Second World War. They were associated with rising incomes and with urbanization, and they were concentrated in a narrow range of products – sugar, wheat (and wheat flour), rice and other 'convenience' foods.[73] These features have persisted; for instance, four-fifths of Nigerian food imports in the 1970s were of wheat, sugar, canned fish, milk and rice. Possibly the rate of increase was faster in the early postwar period than after 1960.[74] Between 1960 and 1978, as was observed in Chapter 4, food imports grew very considerably in absolute terms, but appear to have declined relatively to total imports for several countries (including Ghana, the Ivory Coast and Senegal) and for West Africa as a whole.

Demand for food imports is sometimes attributed to overvalued currencies, which make home-produced foods relatively expensive, and to the effect of food aid on consumer tastes, particularly for wheaten bread. While these explanations are not irrelevant, neither are they sufficient, since the rise in food imports began, and was possibly faster, before currencies were overvalued and food aid became available. Food is imported rather because some foodstuffs for which income-elasticities of demand are high, and which are well adapted to urban styles of life, cannot be supplied locally at costs or in qualities which are competitive with imports. To that extent, food imports are not the aberration or cancerous growth they are often represented to be, but a straightforward result and reflection of economic progress. Policies of deterring these imports in the interest of home production of foodstuffs lower national income even more unequivocally than does import-substitution in manufacturing, since externalities are not postulated to be significant in food-farming.

Table 27 Measurements of domestic resource costs of West African export crops and food crops

		DRC coefficient 1972	*Resource cost ratio 1975*
Export crops			
Cocoa:	Ghana	0.30	
	Ivory Coast	0.36	0.42–0.46 (2 techniques)
Coffee:	Ivory Coast	0.51	0.44–0.58 (2 techniques)
Copra:	Ivory Coast		0.38
Cotton:	Ivory Coast	1.12	0.84–1.03 (2 techniques)
	Mali	0.21	
	Senegal	0.42	0.80
	Togo	0.37 (1977)	
Groundnuts:	Mali	0.23	
	Senegal	0.36	0.48–0.80 (2 techniques)
Palm oil:	Ivory Coast	0.36	0.43
	Nigeria	0.39 (1979)	
Food crops			
Groundnuts:	Nigeria	1.40 (1979)	
Maize:	Ivory Coast		0.81–0.88 (3 techniques)
	Nigeria	1.76 (1979)	
	Senegal		0.80
Millet:	Mali	0.62	
	Nigeria	1.21 (1979)	
	Senegal	0.62	1.27–1.30 (2 techniques)
Rice:	Ivory Coast	1.80 (1975)	1.26–2.99 (11 techniques, av. 1.68)
	Liberia		1.44–1.99 (5 techniques, av. 1.68)
	Mali	0.56 (1976)	0.56–0.99 (7 techniques, av. 0.69)
	Nigeria	2.55 (1979)	
	Senegal	1.02	1.04–2.35 (5 techniques, av. 1.68)
	Sierra Leone		0.69–1.13 (12 techniques, av. 0.89)
Sorghum:	Mali	0.62	
	Nigeria	1.66 (1979)	
	Senegal	0.62	

Sources: DRC coefficients (at official exchange rates) from *Accelerated Development in Sub-Saharan Africa* (Washington DC: World Bank, 1981), p. 65. Resource cost ratios from Scott R. Pearson *et al.*, *Rice in West Africa* (Stanford: Stanford University Press, 1981), Table 12.8, pp. 420–1.

Table 27 shows DRC coefficients, estimated by the World Bank, for several crops in some of the West African countries. The coefficients are calculated at official exchange rates. Equilibrium rates would no doubt have produced lower figures for Ghanaian cocoa in 1972 and for the several Nigerian crops in 1979. The estimates contrast strong comparative advantage in production of nearly all the established export crops (Ivorian cotton is a possible exception) with a more various pattern in food production for domestic markets. Millet and sorghum appear as efficient savers of foreign exchange in Mali and Senegal. So does rice in Mali, and possibly in Senegal, but emphatically not in the Ivory Coast and Nigeria. All five of the food crops shown for Nigeria are apparently costing more to save foreign exchange than the foreign exchange is worth, though millet, and possibly groundnuts, would become efficient if the naira were not overvalued. These coefficients depend, of course, on price relationships, input costs and the technologies in use at the specified dates, and they are not given once and for all. They may be assumed to relate to the crops delivered at major urban markets; domestic resource costs would be much lower, and the coefficients therefore more favourable, for food-crops consumed by their growers or sold in local markets.

The table also shows resource cost ratios (RCRs) for several crops in the Ivory Coast, Liberia, Mali, Senegal and Sierra Leone, according to estimates for 1975 published in a study undertaken for the West Africa Rice Development Association and the Food Research Institute of Stanford University.[75] The RCR is a measurement analogous to the DRC coefficient and can be read in the same way – ratios below unity denote efficient transformation of domestic resources into foreign exchange, ratios above unity imply inefficiency. For rice and some other crops, a range of estimates is given covering various techniques and geographical areas; the averages shown for these ranges are unweighted. It will be seen that the figures agree fairly well with the DRC coefficients produced by the World Bank, with Senegalese millet the only serious inconsistency. Once again, comparative advantage appears in the traditional export crops, and once again it is least in Ivorian cotton. Rice production for the supply of major urban markets, which is the focus of the study cited, appears efficient in Mali and Sierra Leone, but in the Ivory Coast, Liberia and Senegal it can survive only through restrictive trade policy, resulting in income losses from the inefficient use of resources and welfare losses from the higher price of a staple food.

Substitution for imports of wheat and sugar, two crops not shown in the table, is likely to be socially unprofitable nearly everywhere in West Africa. Both crops require irrigation for large-scale production in West African conditions, and capital costs are therefore high. Wheat yields have been disappointing, and heavy subsidies required for the maintenance of production.[76] Sugar production has always been an economic burden in Ghana, though its inefficiency may be partly attributed to the ineptness with which all state enterprises in that country have been established and run.[77] In Nigeria, the Savannah Sugar Project, which was initiated in the early 1970s at Numan in Gongola State, has proved extremely costly both financially and through the negative externalities it has generated (some 50,000 people were dispossessed of their land), and in 1982 it had still to begin production. Like the Ghanaian factories, it will turn out to have been built well ahead of the cane supplies required to keep it fully employed. Sugar refining began in the Ivory Coast in 1974, and production has been projected at 280,000 tonnes by 1985/86, including a surplus of 170,000 tonnes available for export. The World Bank Mission which visited the Ivory Coast in 1975 suggested, on the basis of the data then available, that Ivorian costs could be double the free market price of sugar in the early 1980s.[78] Subsequently, the Ivorian President complained bitterly of the 'levity' with which negotiations for establishing the sugar industry had been conducted, and of faulty financial planning and design and mismanagement: 'All these have resulted in the fact that we are now producing non-competitive sugar, the price of which is higher than the world price.'[79]

While production in disregard of comparative advantage reduces national income, at least in the short term, it also redistributes income in favour of owners of factors of production used in the activities which are protected or subsidized. Policies of substituting home-produced for imported food may be defended on this ground. It is relevant that rice and sugar in the Ivory Coast, rice in Ghana, and sugar and wheat in Nigeria are 'Northern' crops, grown mainly in the less commercialized and developed savanna areas; the same can be said of cotton in the Ivory Coast, the export-crop in which Ivorian comparative advantage is least.

If production in certain areas is to be promoted on the ground of distributive justice, it should be of products in which inefficiency is least pronounced. Often this rule would indicate export-crops like cotton and groundnuts rather than import-substituting crops.

Another formulation of this equity argument is that protection of any crop can be defended as a means of transferring income to the rural areas from the towns. Again, the sacrifice of efficiency would be less if the agricultural protection were discriminative. Indeed, in the general case in West Africa, the rural areas could be assisted without loss of efficiency simply by relieving their most productive activities of punitive taxation.

Even, therefore, if it were accepted in principle that equity demands special treatment for agriculture, either in certain areas or generally, it need not follow that collective self-sufficiency in food would be enhanced as a result.

In practice, it must be added, allowance for equity effects in determining agricultural support is unlikely to be any more exact and coherent, or more immune from political bargaining, than is the compensation for externalities in the protection given manufacturing. For example, government officials and army officers resident in Accra were among the leading beneficiaries of the subsidization of commercial rice production in northern Ghana in the early 1970s; they acquired large tracts of undeveloped land, and combined local labour with the cheap capital and inputs to which their status gave them access. In such a case, redistribution might be thought to be occurring less as a regrettable but necessary trade-off against efficiency than as the primary purpose of economic policy-making.

Security is another argument for self-sufficiency in food. Supplies are postulated to be more secure when obtained domestically than when imported. The implication is that export earnings, adjusted for variations in the price of food imports, are more unstable than are marketed supplies of home-produced food. It is not obvious that this is true, or even that it could possibly be demonstrated (since the statistics of home-market supplies are fragmentary); certainly food production in West Africa cannot be regarded as anything like stable, depending as it does so heavily on weather conditions. Supposing, however, that the proposition were true, it is still not apparent that consumers would be advantaged by less variable food supply, if the food were also more expensive on average and less suited to their tastes.

Possibly food security is less an economic than a strategic consideration. Food imports may be cut off by war, or withheld by the supplying country as a political sanction. It is not apparent that these are real risks in West Africa, or that they should be regarded as significant in countries where agriculture is the principal economic activity of the bulk of the population. A sudden cessation of food imports would make West Africans worse off, but it would not leave them foodless. Is the risk of becoming impoverished in the future a good reason for impoverishing oneself now?

A final argument for contriving the substitution of food, manufactures or anything else for imports is that it 'saves scarce foreign exchange'. Four objections may be made against this somewhat artless claim. First, foreign exchange is not always saved, as has been observed already. Especially in manufacturing, the foreign exchange costs have not uncommonly been found greater for importing inputs and employing foreign factors in home production than for importing the goods for which the home production is substituting. Second, if foreign exchange can be saved, it is best done as cheaply as possible. The saving itself is not evidence of gain. It should not cost more to save a dollar than a dollar is worth to the country saving it. Third, earning more foreign exchange is as good as using less of it; there is ground for preferring the former to the latter wherever it can be done more cheaply.

Fourth, why is foreign exchange 'scarce' – or rather, since all economic goods are by definition scarce, why is foreign exchange regarded as exceptionally scarce? Why is it more important to economize in the use of other countries' resources than to economize in the use of one's own? Evidently, foreign exchange must be exceptionally difficult to earn. The reason might lie in restrictions on West African sales in export markets – but in practice the exports of the region enjoy free or preferential access. Another possibility is that the terms of trade are moving against West Africa – but the estimates presented in Chapter 4 show no such general movement, and indeed the purchasing power of West African exports over imports has expanded enormously since the Second World War.

The real reason for difficulty in earning foreign exchange, and for attributing exceptional scarcity to this resource, lies in domestic policies. Foreign exchange is relatively hard to earn because exporting activities are fiscally penalized – overtly through duties and taxes on profits, often also covertly through statutory monopolies, and sometimes also implicitly through overvaluation of the domestic currency. Domestic currency is easier to earn because many home-market activities are tax-free, and others are explicitly or implicitly subsidized. Ironically, the fiscal discrimination against exporting, which makes foreign exchange relatively scarce, is financially dictated by, among other things, the costs of import-substitution, which is then defended as necessary to 'save scarce foreign exchange'. Hence the last objection to this final justification of policy-contrived transformation of the structure of the economy is that it is the cause of the problem it professes to solve.

Exporting can withstand fiscal discrimination so long as a comparative advantage survives for the producer – so long, that is, as his returns net of imports exceed the returns obtainable from alternative uses of the resources he employs. A high proportion of rent (or surplus over opportunity costs) in the market value of a commodity indicates a great deal of taxable capacity. The penalization of exporting makes itself felt when this capacity is exceeded, and does so through falling volumes of exports. Evidence of falling export volumes in some West African countries in the 1970s, and of deceleration of the rates of growth in others, was mentioned at the beginning of Chapter 4. It was argued then that this decline had been more than offset, for the region as a whole and for its two principal exporting countries, by changes in the composition of

Table 28 Quantities of some West African exports, 1960–2 and 1979–81

	(Annual averages in thousands of metric tons)	
	1960–2	*1979–81*
Ghana: cocoa beans	383	202
Nigeria: cocoa beans	178	189
groundnut products	576	—
palm kernels	399	53
palm oil	155	1
Senegal: groundnut products	557	265
Sierra Leone: palm kernels	58	31
Ivory Coast: coffee beans	149	232
cocoa beans	84	298
logs	763	2,144

Notes: Nigerian, Senegalese and Sierra Leonean figures are for 1960–2 and 1978–80. Sierra Leonean figures are of purchases by the Marketing Board, not exports.

Sources: United Nations, *Yearbooks of International Trade Statistics*; Gill & Duffus, *Cocoa Market Report* (March 1983); Central Bank of Nigeria, *Annual Reports*; Banque Centrale des Etats de l'Afrique de l'Ouest, *Notes d'information et statistiques*; Bank of Sierra Leone, *Annual Reports*.

exports and improvements in the commodity terms of trade. Even so, for several countries including Ghana, Senegal and Sierra Leone, the decline in export volume has become a major source of economic retardation, while the collapse of Nigerian agricultural exports is also evidence of policy-dictated inefficiency in the use of resources.

Table 28 shows the changes in the volumes of some West African recorded exports between the early 1960s and the end of the 1970s. There are also unrecorded, or smuggled, exports, but no plausible allowance for them would remove the impression that production for export of the commodities shown has strongly declined in Ghana, Nigeria, Senegal and Sierra Leone and strongly grown in the Ivory Coast.

The table shows export production of cocoa beans in Ghana, groundnut products in Senegal and palm kernels in Sierra Leone to have been roughly halved between the two periods. The trend in Ghanaian cocoa is particularly remarkable for two reasons. First, it shows the strength of the comparative advantage in that product, for, although cocoa was first singled out for discriminatory taxation in the 1940s, it was not until 1966 that an upward trend in exports was broken, and not until 1974 that the decline became precipitate.[80] Second, the tax regime for cocoa was explicitly connected from 1959 onward with government policies of industrialization.[81] Ghanaian cocoa therefore illustrates with particular force the point that the price paid for premature industrialization consists not only in the higher costs of the products substituting for imports but also in the reduced possibilities of earning by exports. It may be added that other Ghanaian exports – timber, diamonds, gold – have decayed even faster than cocoa since the early 1960s, so that the commodity concentration of the country's exports has actually increased despite the absolute fall in the principal export.

In Nigeria, the export of groundnut products had ceased by 1980, while that of palm produce (and also of cotton, rubber, hides and skins, and timber) was greatly reduced from the levels attained twenty years earlier; among the agricultural exports only cocoa had, at best, been maintained in volume.[82] This decline was accompanied by the rapid emergence of mineral oil exports (see Table 15), and the two movements are connected – not, as the conventional wisdom has it, because the oil revenues allowed the government to 'neglect' agriculture, but rather because expenditure of those revenues raised the opportunity costs of labour and also led, through inflation, to overvaluation of the naira. Some contraction in the resources devoted to agriculture was inevitable in an economy growing as Nigeria's did in the 1970s, but the contraction was accelerated by the greater ability of subsidized non-agricultural activities to withstand increases in labour costs, by the overvaluation of the naira, and by the effects of the discriminatory taxation of farm exports maintained for some thirty years until 1973.[83]

The Ivory Coast provides the contrasting case in Table 28. Between 1960–2 and 1979–81, the export of coffee beans increased by one-half (and would have increased by more but for the quotas imposed by the International Coffee Agreements), while the export of cocoa beans and of logs roughly trebled. In addition, the commodity concentration of exports has diminished through the growth of minor exports, as was shown in Table 14. The effect of these volume increases is vividly brought out when the trend in the income terms of trade of the Ivory Coast is compared with that of Ghana (see Figure 2); while the purchasing power of Ghanaian exports rose by only some 25 per cent between the early 1960s and the late 1970s, that of Ivorian exports quadrupled.

The rapid expansion of the Ivorian economy since about 1950 has been explained as the consequence of a country arriving late (relatively to, for instance, Ghana, Nigeria and Senegal) at the possibilities of venting its surplus resources of land and labour through international trade.[84] Granted that these possibilities were present, they did not have to be exploited. Thus they have not been exploited in Guinea, another late starter with considerable exporting potential.[85]

Ivorian export growth is therefore explained also by policy. Timber exploitation reflects toleration of foreign enterprise, and a willingness to spend public resources in making the forest more accessible to logging. The growth of minor exports such as canned pineapples, rubber, cotton and textiles depends partly on the extension of protection or subsidization to some exporting as well as import-substituting activities. Underlying the contribution of foreign enterprise to the diversification of the economy have been the convertibility of CFA francs into French currency at a fixed rate, freedom in making external transfers, the acceptance of foreign personnel for employment in the Ivory Coast, and the absence of the legal requirements that, in Nigeria and Ghana, have reserved numerous trades to indigenous businesses and have made others permissible only with substantial indigenous participation in ownership.[86] Despite the lessons that are popularly drawn from Africa's colonial experience, and in contrast to the cast of policy in many other African countries, the Ivorian government has usually favoured incorporating the Ivorian economy into that of the wider world.

It is not only through the industries in which non-African enterprise is important that external relationships have continued to be developed. Coffee and cocoa remain the principal Ivorian exports, and they are the products of some half-million agricultural holdings run by Ivorians and other Africans. Producer prices of coffee and cocoa are controlled by a stabilization fund, and considerable fractions of the market value of the crops are withheld from the farmers, as was shown in Table 23. Export agriculture has therefore not been immune from fiscal pressure. Yet enough comparative advantage to the producer has survived for the production of these crops to have continued to trend upward. This result is partly explained by the actual levels at which the producer prices have been fixed. In addition, producers' costs have been held down by heavy public expenditure on roads, which has cheapened transport and extended the area available for profitable cultivation of the crops, and by the free entry accorded workers from neighbouring countries, which has restrained agricultural wages (see Table 13). Finally, although manufacturing has been protected in the Ivory Coast to the detriment of rural consumers, the average level of protection has been moderate, as was noticed above, and non-agricultural output generally may be assumed to have been more competitively and efficiently produced than it would have been if business indigenization had been required.

Ivorian policies do not flow simply, or even mainly, from assessments of economic efficiency or growth possibilities; if they did, they would not encourage production of rice. Like policies elsewhere, they are an outcome of political forces. It was the good fortune of the Ivory Coast, given that its comparative advantage lay so decisively in export agriculture, that the distribution of land holdings was sufficiently unequal for an influential agricultural interest to emerge, and that government since the 1950s has been subjected to that influence.[87] In Ghana, on the other hand, the political influence of farmers did not match their economic predominance, and power was transferred in the late colonial period to a party whose basis of support was mainly urban and which was hostile to the cocoa producing industry. In Nigeria, too, the elite has been urban, industrial, military and bureaucratic in character rather than agricultural; while many compromises have been necessitated by the heterogeneity of Nigerian society, the bias of policy has been toward policy-contrived economic diversification, especially since 1970. Policies of collective self-reliance – and their antithesis, the exploitation of current comparative advantage – therefore rest on particularist interests as well as, or rather than, on ideologies and theories. This observation does not affect the conclusion that it has paid West African countries better to exploit their comparative advantage than to try to change it; it has paid better in economic growth and probably also in satisfying equity objectives.

Exploitation of comparative advantage does not imply that the structure of an economy becomes frozen – that industrialization never occurs, dependence on imported food (or industrial materials, or machines) grows indefinitely, foreign-owned factors of production are never displaced. It means rather that changes in the structure occur when economic actors find it profitable to make them, instead of prematurely, and expensively, when governments find these changes gratifying to their self-esteem or expedient as means of redistributing income and consolidating power.

Summary

Development policies are the politics of development. It follows from the dual commitment which governments in West Africa have made since the 1940s – to the management of economic change and to the delivery, directly or indirectly, of welfare to their populations – that the distribution of incremental income, or of the means of earning it, becomes a principal object of political activity. Hence economic growth on the one hand, and the distribution of income or assets on the other, have been neither separate nor separable issues. The question asked in practice of the 'development effort' has not been how far it succeeds in raising the aggregate of output or material welfare, but whether particular parties, factions, communities and clans are securing the shares in this growing aggregate that they believe due to them.

So long as the resources at official disposal were growing rapidly through external aid and improving terms of trade, it was possible for development policies to emphasize the direct delivery of welfare. When in the 1960s this condition was no longer satisfied, emphasis shifted to the 'trickling down' of welfare from the creation of additional productive capacity. Policies aimed in this direction were then criticized as producing growth 'without development' – a larger GDP, perhaps, but achieved at the expense of national economic autonomy, or without commensurate increase in employment opportunities, or failing withal to relieve poverty, or associated with disequalizing tendencies in society. These criticisms were inspired by populist aspirations, and their authors faced the long-standing populist difficulty[88] of showing their vision of a more egalitarian and harmonious economic order to be somehow compatible with their desire for a continuation, or even acceleration, of economic growth. 'Equity objectives' became voguish in the international agencies in the 1970s, but they lacked strong political appeal for the governments of West Africa, which were in any case without adequate administrative means of pursuing those objectives. It does not follow that distributional considerations were absent from West African development policies; equity aims (of a sort) and equity conflicts had been present all along, and no less conspicuous in the creation of additional productive capacity than in the delivery of welfare directly.

In theory, government management of economic change in areas like West Africa can make it faster and more fruitful and sustained. Two principal aspects of this theoretical case for development policies concern stepping up the rate and controlling the composition of investment, and promoting collective self-sufficiency through the regulation of access to foreign markets, sources of supply and factors of production.

Capital formation was the obvious candidate for the role of prime mover in economic growth. It had been cast in this role by macroeconomic theorizing in the 1940s. It was also the variable that could most readily be reached by international aid. The failure of historical evidence to support the view that economic growth was mainly attributable to capital formation was not allowed to discredit that view; instead, the concept of investment was widened to include human capital, and manpower planning was devised as an adjunct to the functional relationship already postulated.

Disaggregation both of the target GDP and of the 'investment requirement' held to be the means of attaining it provided the detail of central economic plans. Central planning was deemed essential to coordinate investments and so reveal their full social profitability. Setting aside the intrinsic difficulties of this undertaking, it was always a pretence in West Africa because important areas of economic life were omitted, the plans were often deficient in realizable projects, and governments were unwilling or unable to tolerate the restrictions on their freedom of action that adherence to a plan would have imposed.

Planning, raising the investment rate, and increasing government use of resources came to be regarded as much the same thing. As a formula for economic growth, it proved disappointing and sometimes disastrous; Ghana in the 1960s is a notorious, but not unique, illustration. These results cease to be paradoxical if economic growth is recognized to depend less on the supply of investible resources than on the discovery of profitable uses for them. There is much evidence in West Africa of deficiency in the demand for the capital funds that governments have secured, and of consequential under-utilization and low productivity in the assets that have been created. The need to enforce capital flows, either domestically or through inter-governmental aid, is perhaps a sufficient indication that the uses of these funds will be relatively unproductive.

The doctrine that investment governs growth is both unhistorical and, from a strictly economic standpoint, untheoretical. It survives not only because of intellectual conservatism, but also because commercial interests are entrenched in flows of government contracts and international aid.

Policies of collective self-reliance (usually at national level, but conceivably at a regional, or even pan-African, level) are defended as the means of checking the underdevelopment of African countries through their overseas trade and employment of foreign factors. Since African countries do not, in fact, appear to have been retrogressing economically, these policies are alternatively justified as means of removing the structural imbalances or distortions that have been produced by exploitation of current comparative advantage. Unbalanced African economies are held to suffer loss of surplus through their employment of foreign factors, unstable and deteriorating terms of trade through their exportation of primary products, and technical retardation through their failure to build up large-scale manufacturing industries.

The loss of surplus, or remittance of factor earnings abroad, is no measure of the net contribution of foreign factors to national income, and the size of the surplus is in any case a poor guide to the possibilities of economic growth. Unstable earnings are not necessarily undesirable if more stable earnings would also be on average lower. The declining trend which is postulated in the terms of trade of primary-product exporters is not supported by the statistical evidence, and movements in the commodity terms of trade are in any case a misleading indication of the gains from trade. Economic growth through large-scale manufacturing cannot reasonably be said to have been an option available historically to West African countries.

Externalities do (but the infant-industry argument does not) provide some economic ground for policy-contrived transformation of the economic structure, but their fair compensation would require a more exact, flexible and disinterested fiscal administration than is available in West Africa and perhaps anywhere else. The case for policy-contrived transformation is therefore a case for limited and highly selective departures from current comparative advantage. Estimates of DRC and effective protection coefficients for manufacturing industries in Ghana, the Ivory Coast and Nigeria suggest that industrialization has been contrived in fact with scant regard for economic gain or loss.

Governments in West Africa, especially in Nigeria and Ghana, have shown little sensitivity to exposure of the wastes caused by policies of import-substituting industrialization under cover of protection. Indeed, they have asserted the necessity of self-sufficiency in yet another dimension – that of food production. The growth of food imports into West Africa is attributable primarily to rising incomes and urbanization, and the imports are concentrated in commodities in which domestic production for delivery in the towns is seldom competitive. DRC ratios are estimated at well above unity in rice (except in Mali and Sierra Leone) and Nigerian maize, and they would no doubt be high in wheat and sugar. On the other hand, these coefficients appear low for the traditional West African export crops; agriculture is efficient in exporting, but not usually in substituting for imports.

What advantage could there be in reducing dependence on imported food? It might be held to be equitable to encourage production in rural areas, or in particular rural areas, but such production would not necessarily be in substitution for imports. Home-produced supplies of food might be thought more secure, but this belief assumes either that those supplies are more stable than the purchasing power of exports over imports (which may not be so, and, if it were so, would not necessarily be an advantage if home-produced supplies were, on average, more costly); or that prevention of food imports is the only way of avoiding the risk that food imports might be cut off by military or political action (which is nonsense in countries where the bulk of the population makes its living from agriculture).

A final justification of the policy-contrived substitution of manufactures, food or anything else for imports is that it 'saves scarce foreign exchange'. This claim is not always true, and, even where it is true, it does not necessarily imply an economic gain (not if the cost of saving foreign exchange is more than the foreign exchange is worth). This claim raises further the question of why foreign exchange is a resource of exceptional scarcity. In West Africa, the reason is that fiscal and monetary policies have made it much harder to earn than domestic currency. So long as the discrimination of domestic policies against exports is actuated by the objective of substituting home production for imports, that objective is the cause of the problem for which it professes to be a solution.

Declining trends in the export volumes of such commodities as Ghanaian cocoa, Nigerian groundnuts and palm produce, Senegalese groundnut products and Sierra Leonean palm kernels show the effects of export discrimination pushed so far that, in conjunction perhaps with rising real costs, it eliminates or gravely reduces the economic rents accruing to producers of the crops. In the Ivory Coast, on the other hand, while exporting has been heavily taxed, the taxation has not been so great as to prevent rising trends in the production of export commodities; the fiscal burden has been moderated by other policies helping to hold down production costs, notably free access to foreign factors and to foreign exchange itself. Imports and foreign services have not been exceptionally scarce in West Africa when current comparative advantage has been exploited; they are made exceptionally scarce by the promotion of self-reliance in products and factors.

In terms of economic growth, and possibly also of 'equity objectives', exploiting comparative advantage in West Africa has paid better than trying to change it. Comparative advantage may nevertheless be expected to change over time, and the structure of an economy to adapt to those changes. Except where political power has been strongly located in exporting activities, the politics of development in West Africa have not allowed these changes to occur only as and when they become profitable, but have produced the changes prematurely, and therefore often wastefully – redistributing income, certainly, but retarding rather than propelling its growth.

7

Conclusions

The achievements

Like much of the rest of the world, West Africa in the early 1980s was experiencing the strains of economic recession. For most countries in the region, the commodity terms of trade moved adversely after 1979. This movement left badly exposed governments in the Ivory Coast, Liberia, Mauritania, Sierra Leone and Togo which had been borrowing abroad heavily, and imprudently, in earlier years. Debts became insupportable and had to be rescheduled at the inevitable price of surveillance of domestic policies by the International Monetary Fund acting on behalf of the creditors. In Ghana, the change in the terms of trade (more particularly, the relative prices of cocoa and oil) aggravated payments difficulties which were attributable essentially to the government's inability to keep its revenues in line with its expenditures in an economy where official prices diverged wildly from equilibrium values. In April 1983, the official value of the cedi was changed again, but it was found politically desirable to do this not by a devaluation but through a system of massive import surcharges and export bonuses; a little earlier, an attempt had been made in Sierra Leone to legitimize the parallel market in foreign exchange by the introduction of dual official rates.[1] The recession struck Nigeria chiefly through the fall after 1980 in export sales of oil, and a consequential rapid decline in public revenues.[2] Government spending had to be reduced, and an attempt was made in 1982, through intensification of import licensing and in other ways, to reduce by one-half the monthly spending of foreign exchange. The debt service ratio was not high, but Nigeria's ability to raise loans abroad was impeded by arrears in its current external payments, estimated in a range of $3–5 billion at the beginning of 1983.

At least in the cities, these economic reverses heightened social and political tension. There were coups in Upper Volta in

November 1980 and November 1982. There were revolutions (of a sort) in Liberia in April 1980, and in Ghana on the last day of 1981.[3] There was an unsuccessful uprising against the government in The Gambia in July 1981. Industrial relations appear to have worsened in several West African countries; in Freetown, for example, the Sierra Leone Labour Congress organized a general strike against government economic policies in September 1981. Most dramatically of all, the government of Nigeria resolved in January 1983 on the expulsion of aliens illegally resident in the country. One of the greatest of forced migrations followed, some 2.2 million people (according to the official Nigerian estimate) being removed in the course of a few weeks, principally to Ghana and Chad. While this action was probably motivated chiefly by considerations of public security and possible electoral advantage, the background circumstance was the increasing redundancy of labour in the depressed urban economy.

The economic vicissitudes appear to have been accompanied by environmental deterioration, resulting from the coincidence of growing human and animal populations with recurrence of drought in the Sahel, and possibly also from the further destruction of forest for farming in the south, especially in the Ivory Coast.

Irony is nevertheless unintended in the title of this section. The malaise of the early 1980s may be a passing phase. In any event, this book places it in the perspective of a longer period. Measurements have been presented in earlier chapters of considerable advances in West African life and livelihoods in the twenty years following 1960, and these advances continued gains won continuously since 1945. A real GDP series for West Africa as a whole, if one could be computed, would undoubtedly show continuous growth over thirty-five years following the Second World War. The estimates that are available show this growth rate averaging about 4 per cent annually between 1960 and 1979; for Nigeria and the Ivory Coast, economically much the largest countries, the rate is even faster, over 6 and 7 per cent respectively. While the accuracy of GDP estimates must be low (as was stressed in Chapter 2), the impression they give of brisk economic expansion in West Africa is consistent with more reliable statistics relating to physical production, trade and public revenues.

The beneficiaries of the growth in West African production have been predominantly West Africans. Possibly the bulk of the surpluses obtained over the opportunity costs of production has always been retained locally. However that may be, the possibilities of foreigners profiting from West African production have been much reduced in recent times by statutory monopolies of export, heightened competition in the import trade, displacement of expatriate personnel (especially in the public administrations), increased taxation of profits (especially in mining), and the complete or partial nationalization or indigenization of foreign enterprises. It is true that considerable outflows of factor payments and transfers continue from Liberia; but Liberia accounts for little more than 1 per cent of the GDP estimate for all West Africa. It is also true that much of the domestic production of the Ivory Coast (11 per cent in 1978) is lost to Ivorian residents; but a substantial fraction of this loss is workers' remittances to Upper Volta and other West African countries. Allowing for inward transfers (predominantly international aid), it might be guessed that West Africans in the aggregate have been receiving as income in recent years about as much as they have produced as output (albeit the Ivorian record suggests that they could be better off if this were not so).

No doubt the gains from economic growth in West Africa have been unequally distributed, both among countries (of which more below) and among communities and households. Some pressure toward dispersion of gains among communities is exerted by internal politics, especially in Nigeria where a need to hold the country together has been acutely felt since 1966.[4] Reducing inequality among households is another matter; it is hard to see that it lies within the administrative capacity of West African governments, or that there are compelling reasons, whether political or economic, for attempting it. Probably inter-household inequality has increased during the phase of rapid economic growth in West Africa since the Second World War. That would not be inconsistent with historical experience elsewhere.[5] Nor would it be incompatible with general improvement in the absolute levels of household income. Further, it could result as well from some of the poor becoming better off as from the rich getting richer.

Reductions in death rates (especially in the early years of life) and the consequent increases in life expectancy are revealed in countrywide estimates, and therefore these are gains that are extensively enjoyed. Life expectancy in West Africa may have risen by roughly one-half (say, 30 to 45 years) since the 1950s. Though rates of population growth estimated around 3 per cent annually are commonly represented as an appalling social problem, they result from human volitions to procreate and to prolong life, they reflect improvements in the human condition achieved with remarkable speed and generality, and so far they have appeared usually compatible with economic growth in West Africa. Probably the main foundation of the demographic revolution has been preventive medicine, in which case it is a convincing demonstration of the benefits potential in using foreign technology. Even so, mortality rates have fallen faster in the more commercialized and economically buoyant areas, presumably in response to improvements in hygiene, nutrition and curative medicine made possible by higher incomes.

Leisure is another possible dimension of improvement. It could be greatly enlarged in West Africa by reduction in intra-household services. There are no adequate measurements of how far such a reduction has been achieved and it might be thought unlikely to figure prominently in the priorities either of household heads or of governments. On the other hand, some measures are available of the deferment of entry to the labour force (or reduction of participation in economic and intra-household activity) achieved by schooling, and the expansion of formal education has been both an official and a popular priority. The ratio of primary school enrolments to children of primary school age in West Africa as a whole is likely to have doubled (from about 30 to 60 per cent) between 1960 and 1980 and the increase in enrolment ratios in secondary and higher education have been proportionately much greater. There is still a wide divergence in West Africa between the working population and the population of so-called working age, but the divergence has been narrowing rapidly.

Schooling increases leisure, but it is not for this life-enhancing consequence that it has been expanded in West Africa. Education has rather been regarded as financially remunerative, and differentials in earnings support this view, since they depend in part on limitations in labour supply attributable to restricted educational opportunities. Education is also held to be economically productive, which must be true in many instances but is no more generally valid than the assertion that any outlay we choose to label as investment will turn out to be profitable. In addition, the extension of schooling can be seen as an equalizing force, not in the sense that it will make opportunities equal (for innate ability and the standard of educational provision are bound to vary) but in as much as anyone who has not been schooled will be handicapped in a society where literacy is assumed. Finally, perhaps the greatest contribution schooling makes to economic change lies in enlarging mental horizons, raising aspirations and firing ambition – consequences that are not always much appreciated by governments, but arguably provide the soundest case for universal primary education[6] (it follows, incidentally, that such schooling ought not to be designed appropriately to current circumstances, or vocationally relevant to the existing pattern of labour use).[7]

The advances in life expectancy and education have been widely experienced. Only in Guinea, Guinea-Bissau, Mauritania and Sierra Leone do some of the estimates of the current natural increase in population fall below the usual West African range of 2.5–3.5 per cent a year. While estimates of death rates vary widely among West African countries, they have been falling everywhere. The increase in educational enrolments relatively to population has also been widespread. While these ratios remain everywhere low at secondary school level, and low at primary level in the Sahelian countries, they appear to have been rising throughout the region, and the primary rates have now attained high values in Benin, Ghana, Guinea-Bissau, Liberia and Togo as well as in the Ivory Coast and Nigeria (see Table 11).

So, in Ghana, where successive administrations have impeded economic growth ever since the early 1960s, and seemingly achieved reductions in income per head, life expectancy appears nevertheless to have increased and education to have gone on expanding. It was suggested earlier that gains in economic output, the number of population and deferment of entry to the labour force may not be very freely exchangeable. It would have been better, and almost certainly would have been possible, for Ghanaians to have enjoyed faster economic growth along with their other gains, but a slower decline in the child mortality rate or a more limited expansion in schooling would not have done much for the flow of economic goods and services.

Economic growth has been concentrated in Nigeria and the Ivory Coast; their share of the GDP estimate for all West Africa rose from about 50 to 80 per cent between 1960 and 1979. It is also in those countries (along with Ghana) that populations are largest. While two-thirds of West African states fall within the World Bank's category of 'low income' countries, more than two-thirds of West Africans (three-quarters according to Table 1) live in 'middle income' countries. It would be inappropriate in judging the socioeconomic performance of West Africa to regard its sixteen sovereign entities as if they were all of equal significance. Nigeria and the Ivory Coast ought to weigh heavily in any assessment of what has been achieved in West Africa, simply because their residents probably constitute well over one-half of all West Africans. Those residents have included, it will be recalled, large numbers of people attracted from neighbouring countries. In the Ivory Coast, the census of 1975 recorded 1.43 million foreign nationals. In Nigeria, if the official figure of the expulsions is to be trusted, there were at least 2.2 million foreign Africans at the beginning of 1983.

The sources of the achievement

Aggregate income can grow in a country because more human and natural resources are economically utilized, because those resources are used in more productive ways, or because their products become more highly valued in other countries. West African incomes have grown as labour forces have increased through natural population growth and, in some instances, through immigration, as the areas of land under cultivation have been extended, and as mineral and forest resources have been further exploited. They have grown also as labour and land have been drawn from subsistence into commercialized production, and as factors have been combined to produce at lower costs or to satisfy wants more effectively through new products. The growth of West African incomes has also been sustained by improving commodity terms of trade; for the region as a whole (though not for every country), the trend in these terms was mildly favourable between 1960 and 1972 and highly favourable from 1973 to 1979, and there had been an earlier phase of rapid improvement in the first postwar decade.[8]

In a spontaneous economic order, resources are drawn into commercialized use, or put to better uses, by the attraction of market opportunities (including opportunities created by the enterprise of producers). For West African producers, the most attractive opportunities have often been overseas. Labour, land cleared for agriculture and mineral and forest resources could often be most profitably used in supplying for export products whose sales value was frequently much in excess of their costs. These products have included vegetable oils and oilseeds, cocoa, coffee, cotton, rubber, tropical hardwoods, diamonds, crude petroleum, iron ore and other minerals. The same products restricted to domestic markets would have obtained lower prices if they found buyers at all. The same resources excluded from export production would have been unused, uncommercialized, or relatively unremunerative.

Naturally, not all economically utilized resources have been adaptable to exporting activity. Table 5 shows ratios of exports to GDP ranging between about 50 per cent (Liberia in 1980) and only 10 per cent (Ghana in 1977). For West Africa as a whole, this ratio appears as 23 per cent in 1979. (The ratios would, however, be higher for commercialized GDP, and for several countries they would be higher if public policy did not discriminate against exporting.) The export trade has therefore not precluded domestic markets for domestic products. On the contrary, it generated income which helped extend those markets. Further, the profitability of exporting as compared with home-market activities could change over time with changes in relative prices and costs. Resources could therefore shift between these sectors, just as they did among economic activities more particularly divided.

Exporting has nevertheless been the vanguard of growth in West Africa because it has been in exporting that the surpluses over opportunity costs have generally been highest. This feature of West African economic life appears not yet to have changed. The estimates of DRC coefficients shown in Table 27 indicate a continuing marked comparative advantage in the traditional export-crops (and comparative disadvantage in some import-substituting crops). Reference has been made earlier to the enormous economic rents derived from Nigerian oil, most of which is sold abroad; in 1983, the producing companies were allowed costs-plus-profits of $4 per barrel in the tax assessment of equity crude, while participation crude was being sold for export at $30 per barrel.[9] Oil has clearly carried the Nigerian economy forward since the civil war, just as agricultural exports did before.[10] The dependence of Ivorian growth on exports is also unquestioned – least of all by the critics of Ivorian performance. Phases of expansion in smaller countries – Liberia and Mauritania, Togo and The Gambia – have turned on the creation of new exporting industries, increases in export volumes or improving export prices. Conversely, the Ghanaian economy lost momentum when the growth in its exports was checked by policy in the 1960s; forgoing comparative advantage did not discover an alternative source of growth.

Foreign enterprise has frequently participated in the commercialization of West African resources and the reallocations of those resources to more profitable uses. It has engaged in importing incentive goods, securing export outlets, developing new commercial services, creating productive plant, and supplying industrial inputs, technological expertise and skills. As has been said before, the observation that foreign factors are paid for these services tells one nothing about what they contribute to incomes retained in the domestic economy and the growth of those incomes. Improvement in the supplies and productivity of domestic factors can be expected to erode foreign participation as time passes – most easily, perhaps, in ownership of assets, and then in labour skills; least easily in enterprise itself or the making of quintessentially economic decisions. New economic departures in West Africa nevertheless continue to depend often on the cooperation of foreign interests. Petroleum extraction and refining in Nigeria, the great expansion of logging in the Ivory Coast and iron mining in Liberia are notable instances from the period reviewed in this book. Tacitly, at least, governments in West Africa have usually recognized that there is nothing to be gained economically by excluding foreign enterprise, however much they seek to regulate and direct it. Even in Guinea, since 1958 the state most resolutely dedicated to the cause of national autonomy, the economy (and, more particularly, the government) have been held up by American investment in bauxite mining,[11] and by 1982 further foreign interest was being supplicated by the government.

Aid has been another external support of achievements in West Africa since the Second World War. As cheap (and, indeed, usually free) resources, transferred for non-economic reasons and frequently used for non-economic purposes, aid can be expected to have had only a modest direct impact on the growth of domestic production. It does not follow that the aid has been ineffective. Much of it, both in personnel and finance, has been used in establishing and running services in education, public hygiene and preventive and curative medicine. Aid may not have done much for the rate of growth, but its contributions to schooling (especially in the francophone countries) and to reductions in morbidity and mortality rates should not be overlooked. Quantitatively, aid does not appear of great importance for West Africa as a whole (some 2 per cent of 1979 GNP), but the overall measurement masks the considerable dependence on external support of public services in the Sahelian countries, Benin, Liberia and Togo (see Table 26); in those countries, lives would be shorter and ignorance deeper but for aid.

Although foreign enterprise and aid have been crucially important, at least at some times and places, for the achievements under discussion, it would be misleading to think of these forces operating upon an inert or supine mass. Economic opportunities have been exploited, both at home and in overseas markets, principally through the will of West Africans to improve their material conditions of life, even in disregard of the law in those countries where economic policies are economically obstructive; there can be few societies where the materialist ethos burns so brightly as in those of the Guinea coast. West Africans have also sought the survival of their children. Rapid population growth has not occurred involuntarily; it is, in a literal sense, the most popular of causes. The pressure for educational provision has also been strong and widespread, chiefly because of the perceived relationships between educational qualifications, earnings and the chances of occupational mobility. In the early days of development theorizing, it was understood that governments in poor countries would need to inspire their peoples with enthusiasm for material progress.[12] In the West African context, few ideas would appear to have been more detached from reality in this subject where so many received opinions are inconsistent with experience.

The contribution of policy

For which politicians, parties, legislators or administrators is a larger GDP estimate a real ambition? Just who is struggling for development? Politics are concerned more with the distribution of aggregate income than with its growth. Therefore not much should be expected of policies as levers of economic progress.

Policies are sometimes claimed to be means of correcting divergences between private and social profitability, as in the argument for centralized investment planning and contrived industrialization. More commonly, policies create such disparities, deflecting resources from uses where they would be most productively employed, enticing resources into uses where they produce less than they might. Governments could not avoid these consequences, even if they wanted to, because neither their systems of taxation nor their patterns of expenditure are uniform among economic activities – those of West Africa are, in fact, markedly lacking in that characteristic. So, perhaps, exporting is penalized relatively to import-substitution, agriculture relatively to industry, production relatively to trade, informal activities relatively to the formalized, foreign-owned businesses relatively to the indigenous.

West African governments have not been concerned to minimize the distortions unavoidably produced by explicit taxation and the spending of tax proceeds. On the contrary, many of them have enhanced the distortions through deficit financing and consequential inflation, and through the implicit taxes and subsidies created by official pricing of foreign exchange, credit, and a variety of goods deemed essential to consumers or producers.

Income is lost by economic inefficiency; some people, at least, are made materially poorer. There may be compensating advantages. Public administration would presumably be agreed to be necessary, and the collective provision of some services to be desired by everybody; so there has to be taxation. The problems then are to ensure that public administration and services cost no more than they have to, and do not proliferate into areas where they are unnecessary; it was for just such reasons that representative government originated.

Taxes and subsidies (explicit or implicit) may be held to 'improve' the distribution of income among households or localities. There are no objective criteria of improvement, but there is a common presumption that reduction of inequality is desirable. In practice, West African tax systems appear not be progressive, and the ability of West African governments to reduce inter-household inequality through their expenditures has already been questioned.

Distortions that sacrifice income immediately may be a condition of longer-term gains. Chapter 6 was largely taken up with this thesis. It considered first the argument for the enforcement and collectivization of savings. It noted that this argument presupposed an effective demand for capital which, if it existed, would make the enforcement of savings unnecessary; and that, in the absence of such effective demand (which central planning has not, in practice, remedied), the continuing mobilization of resources for investment will have effects on growth which are at best disappointing and at worst immiserizing.

Second, growth has been held to be accelerated in the longer term by policy-contrived changes in the structure of a national economy, creating a more diversified range of activities and specifically a larger manufacturing sector. In so far as this argument turns on shifts in relative prices (specifically, as between primary products and manufactures), it is unsupported by experience and problematic for the future. In so far as it turns on the need to nurture infant industries, it does not show why the costs of this nurturing should be collectively borne. In so far as it turns on the externalities generated by new activities, appropriate compensation for these effects requires more exact, flexible and disinterested fiscal administrations than are found in West Africa. In so far as it turns on the need to save foreign exchange, the argument becomes circular, since the exceptional scarcity of foreign exchange is principally attributable to the policy-contrived changes in the economic structure.

The economic grounds for reallocating resources through policy are therefore less solid and extensive than might be supposed from the prevalence of these interventions in West Africa. The satisfaction of collective wants provides appreciable support, and equity considerations some; growth theorizing, on the other hand, although so often invoked in justification, adds little of significance to the case for manipulating prices, private profitability and resource allocation.

The policy interventions are as prevalent as they are because they affect particular interests. They help govern the distribution of income, assets and economic opportunities; what they do to the rate of growth is often a secondary matter, if it matters in practice at all. Governments exercise this control in order to consolidate and extend, or simply to profit from, their authority. Other actors in the economy seek preferment or protection; they advance their own interpretations of collective wants or equity or social profitability. Any government is exposed to pressures to become the patron of clients, but those of West Africa are particularly vulnerable. They are committed to development, or the delivery directly or indirectly of material welfare, and are judged by their performance of this promise. They preside over economies which, partly because of this commitment, have become heavily politicized, with political favour seen as usually indispensable to, and sometimes a sufficient condition of, commercial success. They govern populations which are ethnically fragmented, and experience sectional conflicts absent from more homogeneous societies. Finally, ministers and officials have frequently sought office, or been propelled into it, precisely in order to control patronage.

The administration of West African economic life therefore shows a marked predisposition in favour of particular rather than general inducements. For instance, crop prices could be raised by scrapping statutory marketing, but this would be a benefit conferred indiscriminately among farmers; the preferred form of support for agriculture is public works projects and subsidization of inputs, assistance which can be distributed selectively.[13] Similarly, the generality of borrowers would be advantaged by the removal of controls on the rates chargeable by institutional lenders, but retaining those controls provides an avenue for patronage. Free competition in the import trade would benefit consumers at large, but particular groups can be benefited when that trade is monopolized.

The entrenchment of particularist interests in policies makes the policies hard to change. Structures of public expenditure become ossified, as was observed in Chapter 5, and can be considerably altered only through a sudden enlargement of the resources at public disposal (whence the urge to inflate). Every government (even a professedly revolutionary government) is trapped by earlier decisions on where and how and on what to spend. By the early 1980s, there was scarcely a West African head of state who did not denounce parastatal bodies as inefficient, wasteful, corrupt and irresponsible. Yet the inference was never drawn that the public corporation was an inappropriate institution in many commercialized activities in West Africa. Instead, the parastatals were adjured to be less inefficient, wasteful, corrupt and irresponsible in the future. Policy-contrived transformation of the structures of African economies was also generally conceded to have been a failure. The explanation, however, was not that the strategy was misconceived, but that it had neglected agriculture. In future, therefore, collective self-reliance must be sought in foodstuffs as well as in manufactures; the mistake was to be rectified by enlarging its scope.

What has policy contributed to the West African achievements if governments have had, in fact, other ends in view than increases in aggregate output and general amelioration in the conditions of life? To a predominant extent, governments have mobilized resources (through taxation, borrowing and foreign aid) to finance the creation and running costs of social and economic infrastructure. In the 1970s, between about three-fifths and three-quarters of total public spending in each West African country was on such functions as education, health care, water supplies, roads, ports, communications and power supply. No doubt the motives behind these expenditures were often less than pure – with facilities distributed according to political criteria, and their costs inflated, sometimes horrendously, through maladministration and because value for money was not usually a primary objective. It does not follow that nobody benefited, or that the emphases were misplaced. Education has been expanded relatively to population in all the West African countries. Health has evidently improved. More purchased energy is consumed.[14] Transport and communications have been developed and cheapened in most of the countries. West Africans wanted more of these services. Governments have tried, however imperfectly, to meet the demand. They have been impelled, not only by the pressure of users or would-be beneficiaries of the services, but also by adventitious increases in the resources at their disposal, arising from inter-governmental aid or, more importantly, from improvements in the terms of trade. Most notably, in Nigeria the ratio of government spending to GDP has risen, and the satisfaction of economic wants has therefore become more collectivized, because incremental national income accrued mainly to the government during the oil boom of 1970–80.

In so far as governments have helped the fuller satisfaction of such wants as those for schooling, water and roads, there has been a rough congruity between policies and demographic and economic improvements. In other respects, policies have often obstructed development – perhaps most so when they have been classed as development policies. The politicization of economic life associated with the governmental commitments to manage economic change and to create material welfare diverts enterprise and skills from the economic to the political arena; not only does political influence provide an alternative way of making commercial gains, but also the other ways tend to be crowded out by the bureaucratic regulation of economic life. The establishment of parastatal bodies as ostensible instruments of development, equipped with coercive powers or monopolistic rights, militates against efficient use of resources; on the one hand, these bodies are immune against competitive pressures, on the other there is constant difficulty in making them publicly accountable, especially where they have been created as the fiefs of influential politicians or to provide wage-employment. Competition, and therefore efficiency, are also diminished by the displacement through legislation or administrative action of foreign (including foreign African) enterprise and labour. The compulsion of governments to harness additional resources can lead to inflation and currency overvaluation, and thence to a progressive curtailment of lawful economic transactions, or even of all exchange relationships; a slogan 'Make war on the economy', used by the military administration in Ghana in the early 1970s, became dreadfully apposite as a characterization of government policy in that country. Most importantly of all, the deliberate forgoing of comparative advantage – especially through the premature shifting of resources from exporting activity to import-substitution – not only sacrifices current income but also prejudices seriously the possibilities of growth unless it is undertaken modestly and with discrimination. Many development policies would be better called policies of redistribution; their consequences are much less uncertain for the distribution of the aggregate income than they are for its growth, and redistribution is the more likely purpose of political activity.

The implications

Governments in West Africa have used national economic gains toward the better satisfaction of collective wants. Their performance of this role has been defective; costs have been higher than they needed to be, benefits have been distributed erratically or inequitably, not all the wants collectively met have required collective satisfaction. Public expenditures on social and economic infrastructure appear nevertheless to have produced positive results in the estimates of mortality and literacy rates and (in most if not all of the countries) in the growing commercialization of economic life.

As active winners of national economic gains, the governments have been much less effective. In some countries, their contribution has surely been negative. This outcome is unsurprising. Politicians and public administrators do not usually possess special aptitude for perceiving and exploiting economic opportunities. They may well suffer special disabilities, if profit-motivated responses to changing opportunities are thought to threaten the bases of political support or to be disruptive of orderly administration. Governments *qua* governments are not actuated by the profit motive. Usually they count this to their credit. They want larger GDPs, but it is more important to stay in office.

A government may therefore become in practice an interest

opposed to economic growth. This opposition interest can develop through political pressures on the government to conserve or extend public payrolls and to maintain or increase the subsidization of particular groups of private producers or consumers. It does not necessarily extinguish growth, but it is an incubus against which the forces making for growth may have to struggle. It produces a legitimizing ideology, in which socialist and populist ideals, distributive justice, the necessity of central planning, national self-esteem and escape from dependence on foreigners characteristically feature.

So commonplace are the economically conservative or reactionary consequences of politics that one might well ask how governments can possibly avoid becoming an interest opposed to economic growth. This question lay at the heart of classical political economy in the early nineteenth century, and was answered by the prescriptions of free trade, freedom of contract, low taxation, and the preservation of profit inducements. By Marx's account, government became no more than a committee for managing the common affairs of the whole bourgeoisie – a function which served not only bourgeois interests but also the Marxist imperative that the forces of production be developed as fully and rapidly as possible. In Schumpeter's model of capitalist development in the early twentieth century, economic policies (and, indeed, the socio-psychological climate) are implicitly regarded as so far favourable to entrepreneurship or economic innovations as to make unnecessary any analysis of government action.[15]

As those citations suggest, it matters less for economic growth that the groundrules should be known than that there should be convergence between the interests which those rules would serve and the locus of political power. In the West African case, growth typically requires the political predominance of exporting interests. The colonial order broadly satisfied this requirement, albeit with considerable administrative misgivings;[16] denied external subvention and an autonomous monetary system, and possessed of a very narrow tax base, the authorities encouraged exporting in order to raise revenue from import duties. In British West Africa, this consonance between policy and comparative advantage began to fade in the last phase of colonialism, as was explained in the discussion of the marketing boards in Chapter 5. The discords became louder after independence – notably in Guinea and Mali, as well as in Ghana, Nigeria and Sierra Leone.

The Ivory Coast evolved differently. Chapter 6 was concluded by citing Bates's suggestion that the tenor of Ivorian policies is explained by the strong political influence of an agricultural interest arising from the unequal distribution of landholdings. The fiscal squeeze on export crops has not been so great as to check their growth. Costs have been held down, and incentives held up, by an elastic supply of cheap immigrant labour, moderation in industrial protection, unlicensed imports and public spending on transport improvements. Care was taken to maintain the monetary arrangements with France that inhibited deficit financing and therefore inflation, and to continue preferential trading and concessional financial relationships with France and later with the EEC. These policies were permissive of rapid economic growth sustained over some thirty years. The effects of a different locus of power, hostile to agriculture, are dramatically apparent in the contrasting fortunes of Ghana, or of Guinea.

Nigeria's economic expansion since the 1960s is otherwise explained. Agricultural exports have been mostly suppressed, first by fiscal pressure and later by rising opportunity costs. Growth has depended overwhelmingly on oil exports produced by foreign enterprises, and on increased taxation of and public shareholdings in those enterprises. Nigeria has grown on the basis of local appropriation of the economic rent created by mining a natural resource.

The fortunes of the initial users of this rent (politicians, public employees, the army, government contractors) are linked with maintaining oil production. It is a powerful connection, but not necessarily strong enough to prevent maladroit policies. The official classes in Ghana were scarcely less dependent on cocoa production, but they garrotted the industry nonetheless. It would pay Nigeria economically if the oil multinationals had a stronger voice in Nigerian affairs; failing sufficient voice, they have the alternative of exiting. In any event, productive capacity in oil will eventually run down – at some point in the 1990s on the basis of the reserves proven and the rate of extraction current in 1983: perhaps a decade or so later, if further exploration for oil is made financially worthwhile.

If economic growth is to continue, a need will soon become pressing in Nigeria, as it has long been pressing in Ghana, for resources to be shifted out of uses in which they are socially unprofitable, either relatively to alternative uses or even absolutely. No one acquainted with West Africa would doubt that these economic losers include the detailed administrative regulation of economic life and many forms of public enterprise. There would be less agreement that they include many protected import-substituting activities. Yet the externalities argument for this protection, which economically is the only one worth taking seriously, carries an enormous burden if it is used to justify production costs two or three times as high as import costs, as will be the case with Nigerian steel.

Much will depend on revival of world trade in energy, minerals and tropical crops. West Africa retains pronounced comparative advantage in many of its traditional agricultural exports, and the land and labour are surely available for greater export production in Ghana, Guinea and Nigeria. The Ivory Coast is expected to become an oil exporter by 1985. There are enormous reserves of natural gas in the Niger delta; most of the gas extracted in association with oil has presently to be burned away. Large reserves of iron and other mineral ores await exploitation in Guinea, Senegal, Ghana and elsewhere.

To take full advantage of whatever opportunities international trade may provide in the future would require reversal of the anti-growth policies which have become increasingly characteristic of many West African countries since the 1950s. As experience in Ghana has amply demonstrated in practice, it is difficult in the extreme to reverse these policies, because of the interests which are entrenched in their redistributive consequences. Also, there are other things in life, and more important things, than increasing the flow of exchangeable goods and services; this book has made some allusion to them. Even so, the opportunities for economic growth which are presented by natural resources and enterprising populations should not be lightly forgone in a region as materially poor as West Africa.

NOTES AND REFERENCES

Chapter 1: West Africa as an Economic Region

1. W. B. Morgan and J. C. Pugh, *West Africa* (London: Methuen, 1969), pp. xviii–xix, 724–5.
2. The Northern Cameroons became part of the Northern Region of Nigeria in 1961, while the Southern Cameroons became the Western State of the Cameroon Republic.
3. Morgan and Pugh, op. cit., pp. xxiv–xxv.
4. In practice, the basic distinctions made in national accounting are more conventional than logical. Thus, final products are distinguished from intermediate products mainly by designating households, governments and foreigners as final buyers. Fuel purchased by a household or security of property maintained by the police thus become final products, while the electricity bill of an enterprise or the costs of its own security service count as expenditures on intermediate goods. The boundary between exchangeable and other goods and services is also arbitrarily drawn. For example, since some households hire help for cooking and cleaning, laundering and the care of children, it might be argued that most services rendered without payment among members of a household are in principle exchangeable; intra-household services are nevertheless not normally included in estimation of the GNP. Even the distinction between Gross and Net National Product has appeared somewhat capricious since the concept of investment was extended from directly income-earning assets to include social overhead capital in the 1930s and human capital in the 1960s; there is a sense in which all the GNP is used to maintain or to increase stocks of capital.
5. An exception to the usual official reticence is the *National Accounts of Sierra Leone* (Freetown: Central Statistics Office, 1967), p. 40, where the sectors of GNP were classified in three categories of reliability, each accounting for about one-third of the total: (a) 'quite reliable (under 5 per cent margin of error)', (b) 'reasonably reliable (5–15 per cent margin of error)'; and (c) 'subject to a wider margin of uncertainty'. A revision of the Ghanaian GNP estimate for the period 1965–8 reached the conclusion that it had been overstated by about 15 per cent: see M. S. Singal and J. D. N. Nartey, *Sources and Methods of Estimation of National Income at Current Prices in Ghana* (Accra: Central Bureau of Statistics, 1971). An estimate of an 8 per cent error in the Liberian GDP *excluding subsistence output* in the mid-1960s was thought by later investigators to be 'very optimistic': see *The Tax System of Liberia: Report of the Tax Mission* (New York & London: Columbia University Press, 1970), p. 37. Revisions in 1979 of the Nigerian national accounts since 1973 raised the GDP total by more than one-third in some years: compare the estimates published in IMF, *International Financial Statistics* before and in 1981.
6. D. W. Blades, 'What do we know about levels and growth of output in developing countries? A critical analysis with special reference to Africa', in *Economic Growth and Resources*, vol. 2, ed. R. C. O. Matthews (London: Macmillan for the International Economic Association, 1980). It may be added that Kuznets

thought it reasonable to assume an average margin of error of about 10 per cent in the GNP estimates of the United States during the interwar period, while for the sectoral components of these estimates, considered separately, margins of error were estimated in a range from 7.5 to about 40 per cent: Simon Kuznets, *National Income and its Composition 1919–1938* (New York: National Bureau of Economic Research, 1941), vol. II, ch. 12.

7. Irving B. Kravis, Alan W. Heston and Robert Summers, 'Real GDP *per capita* for more than one hundred countries', *Economic Journal* 88 (1978).
8. See the discussion of Kenyan inequalities in Arthur Hazlewood, *The Economy of Kenya: the Kenyatta Era* (Oxford: Oxford University Press, 1979), pp. 190–7.
9. References include Morgan and Pugh, op cit., p. 447; Jean Rouch, *Migrations au Ghana . . . Enquête 1953–1955* (Paris: Société des Africanistes, 1956), p. 100; International Labour Office, *African Labour Survey* (Geneva: ILO, 1962), p. 133; B. Gil, 'Immigration into Ghana and its Contribution to Skill', in United Nations, *Proceedings of the World Population Conference, Belgrade, 1965*, vol. IV (New York: UN, 1967), p. 202; *Situation Economique de la Côte d'Ivoire 1960* (Abidjan: Ministère des Finances, des Affaires Economiques et du Plan, Direction de la Statistique, 1961); R. Mansell Prothero, 'Migratory Labour from North-Western Nigeria', *Africa* 27 (1957).
10. Samir Amin (ed.), *Modern Migrations in Western Africa* (London: Oxford University Press for the International African Institute, 1974), p. 74.
11. The census data for the nine countries are analysed with respect to migration in K. C. Zachariah and Julien Condé, *Migration in West Africa: Demographic Aspects* (Oxford: Oxford University Press for the World Bank, 1981).
12. Gil, op. cit., p. 203.
13. Zachariah and Condé, op. cit., p. 10.

Chapter 2: Economic Structures

1. Estimates of margins of error in African GDP statistics are made in D. W. Blades, op. cit.
2. The price of marketed units of a product could reasonably be equated with that of unmarketed units if the latter were few relatively to the former. Where this condition does not hold, as in most subsistence output in West Africa, the imputation of price must be arbitrary since market prices would obviously change if relatively large quantities were diverted from subsistence to sale. 'Subsistence', it may be added, is a misleading term if it is taken to imply a bare minimum of existence; most economically active persons engage both in subsistence and exchange activities, in West Africa as elsewhere. Alternative terms are non-marketed output, own-account product and auto-consumption.
3. Thus, it was estimated in the mid-1950s that about two-thirds of the total value of trade between AOF and British West Africa was contraband; the value of diamond smuggling from Sierra Leone in the late 1950s was believed to be almost as great as that of the country's legal exports; and cocoa smuggling from Ghana in the late 1970s was alleged to be as great as 50,000 tonnes annually.
4. For further discussion, see Douglas Rimmer, 'Official Statistics', in Margaret Peil *et al.*, *Social Science Research Methods: an African Handbook* (London: Hodder & Stoughton, 1982).
5. P. N. C. Okigbo, *Nigerian National Accounts 1950–57* (Enugu: Federal Ministry of Economic Development, 1962). Because of changes in methods, these

estimates are not strictly comparable with a series maintained from 1958/59 by the Federal Office of Statistics. The latter in turn are inconsistent with a series running from 1973/74 that was first presented in 1979.

6. *Policies for Industrial Progress in Developing Countries*, ed. John Cody, Helen Hughes and David Wall (Oxford: Oxford University Press for the World Bank, 1980), p. 21.
7. *World Development Report 1980* (Oxford: Oxford University Press for the World Bank, 1980), Annex Table 3.
8. As in Cody, Hughes and Wall, *Policies for Industrial Progress*, pp. 21–34.
9. *Ivory Coast: the Challenge of Success* (Baltimore & London: Johns Hopkins University Press for the World Bank, 1978), pp. 35–6.
10. *A Study of Contemporary Ghana*, vol. I, ed. Walter Birmingham, I. Neustadt and E. N. Omaboe (London: George Allen & Unwin, 1966), pp. 62–6.
11. Wolfgang F. Stolper, *Planning without Facts* (Cambridge, Mass.: Harvard University Press, 1966), Appx by Nicholas G. Carter on 'An Input-Output Analysis of the Nigerian Economy, 1959–60'.
12. Tony Killick, *Development Economics in Action: A Study of Economic Policies in Ghana* (London: Heinemann, 1978), p. 169; *Five-Year Development Plan 1975/76–1979/80* (Accra: Ministry of Economic Planning, 1977), Part I, p. 2.
13. Private consumption fell from 67 per cent of the Nigerian GDP estimate for 1973/74 to 54 per cent in 1974/75, while the export surplus grew from 7 to 21 per cent of the GDP.
14. Douglas Rimmer, 'Development in Nigeria: an Overview', in *The Political Economy of Income Distribution in Nigeria*, ed. Henry Bienen and V. P. Diejomaoh (New York: Holmes & Meier, 1981).
15. *International Financial Statistics Yearbook 1980* (Washington DC: International Monetary Fund, 1980), p. 319.
16. The sixty 'middle-income' countries are those, ranging from Kenya to Spain, with 1979 estimates of GNP per head exceeding $370. Their weighted average uses of GDP in 1979 were as follows (with the equivalent West African figure in brackets): private consumption, 62 per cent (62); government consumption, 13 per cent (11); gross domestic investment, 26 per cent (28); resource balance, −1 per cent (−1); gross domestic savings, 25 per cent (27) (*World Development Report 1981* (Oxford: Oxford University Press for the World Bank, 1981), Annex Table 5).
17. Okigbo, *Nigerian National Accounts*, Table IV.6, p. 25; *Economic Survey 1955* (Accra: Office of the Government Statistician, 1956), Appx I, Table 8.
18. *Yearbook of National Accounts Statistics 1979* (New York: UN, 1980), vol. I, pp. 477, 1187, 1356.
19. The estimates appeared in the annual *Economic Surveys* published by the Government Statistician's Office, Accra; see especially those for 1957 and 1958.
20. Robert W. Clower, George Dalton, Mitchell Harwitz and A. A. Walters, *Growth without Development: An Economic Survey of Liberia* (Evanston, Illinois: Northwestern University Press, 1966), Appx 2A, pp. 41–61.
21. By 1980, government income from property (i.e. from exploitation of oil deposits) probably constituted over one-quarter of the Nigerian GDP estimate.
22. Such a conception was approached in Liberia in the 1960s by the exclusion from the GNP of the entire compensation of non-African employees. It was commented (in *The Tax System of Liberia*, op. cit., Appx to ch. II) that, for

consistency, the compensation of African but non-Liberian employees and the depreciation of foreign-owned capital should also have been excluded. But it is possible that the exclusion of non-African employees' earnings arose simply from a belief (or convenient assumption) that these employees did remit abroad all their salaries, while their living expenses in Liberia were met by the companies that employed them.

23. The aggregate of workers' remittances sent through official channels from The Gambia, Ghana, the Ivory Coast, Mali, Senegal, Togo and Upper Volta in the period 1970–4 has been estimated at $705 million (including $502 million from the Ivory Coast), and that of workers' remittances received by the same countries in the same period as $233 million (including $137 million by Upper Volta). The discrepancy between payments and receipts results from the presence of migrant workers from other countries (both African and non-African) and from recording deficiencies (Zachariah and Condé, op. cit., pp. 52–3).
24. Available estimates of national and national disposable incomes appear in the United Nations *Monthly Bulletin of Statistics* and *Yearbook of National Accounts Statistics*.
25. *Nigeria: Options for Long-Term Development – Report of a Mission sent to Nigeria by the World Bank* (Baltimore & London: Johns Hopkins University Press, 1974), Statistical Annex, Table 8, p. 209.
26. The estimates for 1960 and 1964 appear in R. M. Barkay, *Domestic and National Product for Liberia 1964: Preliminary Estimates* (Monrovia: Office of National Planning, 1966), and those for 1968 in *The Tax System of Liberia*, pp. 50–2. On the exclusion of non-African wages and salaries, see note 22 above.
27. For examples of such procedures in Nigeria, see A. R. Prest and I. G. Stewart, *National Income of Nigeria 1950/1951* (London: HMSO, Colonial Research Studies no. 11, 1953), p. 86; Adebayo Adedeji, *Nigerian Federal Finance* (London: Hutchinson Educational, 1969), Table 2.6, p. 40, and Appx D, pp. 290–2; and V. P. Diejomaoh and E. C. Anusionwu, 'The Structure of Income Inequality in Nigeria: a Macro Analysis', in Bienen and Diejomaoh (eds), op. cit.
28. A point on the graph of the frequency function would relate any given size of income to the percentage frequency of the population with that income. A point on the Lorenz curve would indicate what percentage of total income accrues to any specified percentage of the population. In the latter case, absolute equality would be represented by a 45° line (any percentage of the population receiving the same percentage of the total income). The Gini coefficient is the ratio of the area between the Lorenz curve and the 45° line to the total area beneath the 45° line. A coefficient of zero represents absolute equality, a coefficient of unity absolute inequality. For a concise discussion, see Richard Szal and Sherman Robinson, 'Measuring Income Inequality', Appx I in *Income Distribution and Growth in the Less-Developed Countries*, ed. Charles R. Frank Jr and Richard C. Webb (Washington DC: the Brookings Institution, 1977).
29. Estimates of the size-distribution of income in several West African countries, mostly for 1959 or 1960, were first presented in Christian Morrisson, *La Répartition des Revenus dans les Pays du Tiers Monde* (Paris: Editions Cujas, 1968), and in several mimeographed papers by Irma Adelman and Cynthia Taft Morris. Estimates for Sierra Leone have been drawn from urban household

surveys made in the late 1960s. Morrisson made further estimates for the Ivory Coast in 1970 in an unpublished World Bank paper. The more readily accessible sources for these estimates are Felix Paukert, 'Income Distribution at Different Levels of Development: A Survey of Evidence', *International Labour Review* 108 (1973); Shail Jain, *Size Distribution of Income: a Compilation of Data* (Baltimore: Johns Hopkins University Press for the World Bank, 1975); and Montek S. Ahluwalia, 'Income Inequality: some Dimensions of the Problem', in Hollis Chenery *et al.*, *Redistribution with Growth* (London: Oxford University Press, 1974). Later (1973/74) estimates for the Ivory Coast appear in *Ivory Coast: the Challenge of Success*, pp. 130–6 and 380–1, and have been criticized by Eddy Lee, 'Export-Led Rural Development: the Ivory Coast', *Development and Change* 11 (1980), pp. 625–7. On Nigeria, see Diejomaoh and Anusionwu, 'Structure of Income Inequality in Nigeria', loc. cit., pp. 109–16.

30. Thus, visiting economists surveying the Liberian economy in 1960 reached a rough conclusion that 'fewer than 5 per cent of the total number of income-receiving units in the economy receive more than 90 per cent of total domestic income in money and kind' (Clower *et al.*, *Growth without Development*, p. 65).
31. Diejomaoh and Anusionwu, 'Structure of Income Inequality in Nigeria', loc. cit., p. 115.
32. This share becomes one-fifth in 1973/74, according to *Ivory Coast: the Challenge of Success*, but the World Bank Mission appears to have been estimating inter-departmental rather than interpersonal distribution, as is pointed out by Lee, 'Export-Led Rural Development'.
33. A case in point is the data for Sierra Leone given in Ahluwalia, 'Income Inequality', Table I.1, p. 8. The lowest 40 per cent of households are shown as receiving 9.6 per cent of total income, the next 40 per cent as receiving 22.4 per cent, and the top 20 per cent as receiving 68.0 per cent. Ahluwalia gives his source as the *Africa Research Bulletin* 5, no. 1 (1968). This bulletin is a newspaper-abstracting periodical. Its own source was the Freetown *Daily Mail*, which in turn had drawn its information from the advance report of the *Household Survey of the Western Province, November 1966–January 1968* (Freetown: Central Statistics Office, 1967). The item in the Africa Research Bulletin reports that 18 per cent of the surveyed households in the Western Province received less than 20 leones per month, 38 per cent between 20 and 40 leones, and 80 per cent over 100 leones. There are two errors in this report: the figure of 80 per cent should read 8 per cent, and the information given relates to Freetown (containing 68 per cent of the surveyed households) not to the Western Province as a whole. More important are two other considerations. First, it is clearly impossible to compute from this information the figures given by Ahluwalia. Second, so far as the *Household Survey of the Western Province* has been made use of by Ahluwalia, it relates to the capital city and its immediate neighbourhood and cannot be considered representative of Sierra Leone as a whole (for instance, only 7 per cent of the surveyed households were classed as rural). It may be added that the survey measured only money-income; this is presumably the explanation of why some households were recorded as receiving no income at all.
34. See the review of findings for Taiwan, Pakistan and Columbia in Gary S. Fields, *Poverty, Inequality and Development* (Cambridge: Cambridge University Press, 1980), pp. 111–14.

35. See Albert O. Hirschman, 'The Changing Tolerance for Income Inequality in the Course of Economic Development', *Quarterly Journal of Economics* 87 (1973). For a review of empirical evidence of perceptions and evaluations of inequality in a West African country, see Donald G. Morrison, 'Inequalities of Social Rewards: Realities and Perceptions in Nigeria', in Bienen & Diejomaoh (eds), op. cit.
36. For a fuller discussion, see Douglas Rimmer, *Macromancy: the Ideology of 'Development Economics'* (London: Institute of Economic Affairs, Hobart Paper 55, 1973).
37. Oskar Morgenstern, *On the Accuracy of Economic Observations* (Princeton: Princeton University Press, 2nd edn, 1963), p. 286.

Chapter 3: Population and Labour Force

1. Those are the adjustments made in US Bureau of the Census, *Country Demographic Profiles – Ghana* (Washington DC: Bureau of the Census, 1977).
2. On the Nigerian censuses of 1962 and 1963, see R. K. Udo, 'Population and Politics in Nigeria', in *The Population of Tropical Africa*, ed. J. C. Caldwell and C. Okonjo (London. Longman, 1968), and S. A. Aluko, 'How many Nigerians?', *Journal of Modern African Studies* 3 (1965).
3. Some account of public reaction to the 1973 census results is given in L. C. Dare, 'Nigerian Military Governments and the Quest for Legitimacy, January 1966–July 1975', *Nigerian Journal of Economic and Social Studies* 17 (1975), pp. 107–12.
4. See *Demographic Yearbook 1979* (New York: UN, 1980), Table 3.
5. *Accelerated Development in Sub-Saharan Africa* (Washington DC: World Bank, 1981), Annex Table 33.
6. Ibid., Annex Table 34.
7. *Demographic Yearbook 1979*, Table 22.
8. See Zachariah and Condé, op. cit., p. 79.
9. See Josef Gugler and William G. Flanagan, *Urbanization and Social Change in West Africa* (Cambridge: Cambridge University Press, 1978), Table 2.2, p. 39.
10. The demographic surveys conducted in Benin and Liberia show the following contrasts:

	Benin 1961		Liberia 1970	
	Urban	Rural	Urban	Rural
Infant mortality rate	45.5	116.6	91.4	158.6
Child mortality rate	19.6	47.8	13.1	19.8
Death rate, all ages	11.9	27.5	11.6	18.1

(*Source: Demographic Yearbook 1979*, Table 20.)

11. Ibid., Table 8.
12. Ibid., Table 7.
13. The Nigerian estimate is derived from *Outline of the Fourth National Development Plan 1981–85* (Lagos: Federal Ministry of Planning, n.d.), p. 88. The estimates for other countries come from *Year Book of Labour Statistics* (Geneva: ILO), various issues, Table 1.
14. K. C. Doctor and H. Gallis, 'Size and Characteristics of Wage Employment in Africa: some statistical estimates', *International Labour Review* 93 (1966).
15. Heather Joshi, Harold Lubell and Jean Mouly, *Abidjan: Urban Development and Employment in the Ivory Coast* (Geneva: ILO, 1976), Table 24, p. 103.

16. *Outline of the Fourth National Development Plan*, p. 88.
17. It would also require the disappearance of aged and infirm people. If deceleration in population growth were accompanied by increasing longevity, it would eventually raise rather than lower the dependency ratio, because of an increasing proportion of aged people.
18. Albert O. Hirschman, *The Strategy of Economic Development* (New Haven: Yale University Press, 1958), p. 181.
19. See the discussion of this point in A. G. Hopkins, *An Economic History of West Africa* (London: Longman, 1973), pp. 76–7: 'Underpopulation was critical in preventing market growth because it encouraged extensive cultivation, favoured dispersed settlement and generated strong tendencies towards local self-sufficiency.'
20. David Morawetz, *Twenty-Five Years of Economic Development 1950 to 1975* (Washington DC: The World Bank, 1977), Statistical Appx, Tables A1 and A6.
21. Simon Kuznets, *Modern Economic Growth: Rate, Structure and Spread* (New Haven: Yale University Press, 1966), pp. 63–8.
22. *Accelerated Development in Sub-Saharan Africa*, pp. 112–13.
23. Ibid., Table 8.1, p. 112, and Annex Table 33.
24. The estimates for 1948 and 1960 are given by J. C. Caldwell in *A Study of Contemporary Ghana*, vol. II (London: George Allen & Unwin, 1967), ed. Walter Birmingham, I. Neustadt and E. N. Omaboe, p. 90; Caldwell also made a rough estimate of 28 years for 1921 and a projection of 53 years for 1975 (ibid., pp. 93 and 104). The estimate for 1970 appears in US Bureau of the Census, *Country Demographic Profiles – Ghana*, and corresponds with the World Bank estimate for 1979.
25. *The Future Growth of World Population* (New York: UN, 1958), pp. 3–5.
26. P. T. Bauer, *Dissent on Development* (London: Weidenfeld & Nicolson, 1971), pp. 63–4.
27. Both water supply and schooling are included in estimates of the GDP, but valued at the costs of providing them, without regard for the values created by the consequential increases in leisure for women and children.
28. UNESCO, *Statistical Yearbooks* 1972 (Tables 3.2, 3.4 and 4.1) and 1980 (Tables 3.4, 3.7 and 3.11).
29. *A Summary of Current Education Statistics in Nigeria 1978–79* (Lagos: Federal Ministry of Education, 1980).
30. See Morris David Morris, *Measuring the Condition of the World's Poor* (New York: Pergamon Press for the Overseas Development Council, 1979), where a Physical Quality of Life Index is proposed as an alternative to GNP per head, based on the infant mortality rate, life expectancy at the age of one, and the literacy rate. According to this index, the wellbeing of national populations (and other social groups) tends to converge over time because there is scarcely anywhere left to go when an infant mortality rate of 8 per thousand, life expectancy of nearly seventy-five years, and a literacy rate of 99 per cent have been achieved (as in Sweden).
31. See the data for The Gambia, Ghana, the Ivory Coast, Liberia, Mali, Senegal, Sierra Leone, Togo and Upper Volta in Zachariah and Condé, op. cit., pp. 79–97, and estimates for the Sahelian countries cited in *Urban Growth and Economic Development in the Sahel* (Washington DC: World Bank Staff Working Paper no. 315, 1979), pp. 18–19. In Nigeria, the contribution of in-

migration to urban population growth is probably no less important, but the movement is not heavily concentrated in a single city.

32. On the migrants' own views, the principal findings are those obtained by a large survey in Ghana in 1963 and summarized by J. C. Caldwell in *A Study of Contemporary Ghana*, vol. II, pp. 136–44. See also the findings of a survey of in-migrants in Lagos in 1972, cited in Olanrewaju J. Fapohunda and Harold Lubell, *Lagos: Urban Development and Employment* (Geneva: ILO, 1978), Table 2.7, p. 35.
33. International Labour Office, *Report to the Government of Ghana on Questions of Wage Policy*, by J. E. Isaac (Geneva: ILO, 1962).
34. Peter Kilby, *Industrialization in an Open Economy: Nigeria 1945–1966* (London: Cambridge University Press, 1969), pp. 280–1.
35. W. Arthur Lewis, *Reflections on Nigeria's Economic Growth* (Paris: Development Centre of the OECD, 1967), p. 42.
36. These costs might be construed to include the costs of obtaining information as well as travelling expenses.
37. For further discussion of the supply price, see J. B. Knight, 'Rural-Urban Income Comparisons and Migration in Ghana', *Bulletin of the Oxford University Institute of Economics and Statistics* 34 (1972): 199–228.
38. See the somewhat adventurous comparisons of average formal and informal sector incomes in *First Things First: Meeting the Basic Needs of the People of Nigeria* (Addis Ababa: ILO, 1981), pp. 218–19.
39. Keith Hart, 'Informal Income Opportunities and Urban Employment in Ghana', *Journal of Modern African Studies* 11 (1973): 61–89, and 'Migration and the Opportunity Structure: a Ghanaian case-study', in *Modern Migrations in Western Africa* (London: Oxford University Press for the International African Institute, 1974), ed. Samir Amin, pp. 321–39; Margaret Peil, 'West African Urban Craftsmen', *Journal of Developing Areas* 14 (1979): 3–22.
40. In 1981, such supplements were said to be worth ₦10–30 in relation to a Nigerian minimum wage of ₦125 per month.
41. Kilby, *Industrialization in an Open Economy*, p. 278n.
42. Guy Pfefferman, *Industrial Labor in the Republic of Senegal* (New York: Praeger, 1968), pp. 160–70.
43. Joshi *et al.*, *Abidjan*, pp. 39–43, citing Ministère des Finances, des Affaires Economiques et du Plan, *Etude socio-économique de la zone urbaine d'Abidjan* (Paris: 1966–8).
44. Ibid., pp. 36–7.
45. John Maynard Keynes, *The General Theory of Employment, Interest and Money* (London: Macmillan, 1936), p. 161.
46. Michael P. Todaro's model of rural-urban migration brings out this point; see, for instance, his 'Income Expectations, Rural-Urban Migration and Employment in Africa', *International Labour Review* 104 (1971): 387–413.
47. The ability to give hospitality and other support to relatives and friends might, however, be interpreted as part of a standard of living.
48. *Second National Development Plan 1970–74* (Lagos: Federal Ministry of Information, 1970), p. 72.
49. Calculated from data in *Ivory Coast: the Challenge of Success*, op. cit., Table 6.3, p. 130.
50. *Economic Survey of Liberia 1974* (Monrovia: Ministry of Planning and

Economic Affairs, 1975), Table 3.4, p. 39.

51. See Olufemi Fajana, 'Aspects of Income Distribution in the Nigerian Urban Sector', in Bienen and Diejomaoh (eds), op. cit., Table 6.4, p. 208.
52. Keith Hinchliffe, 'Education, Individual Earnings and Earnings Distribution', *Journal of Development Studies* 11 (1974/75): 149–61.
53. O. Aboyade, *Incomes Profile* (Ibadan: University of Ibadan Press, 1973), cited in V. P. Diejomaoh and E. C. Anusionwu, 'Education and Income Distribution in Nigeria', in Bienen and Diejomaoh (eds), op. cit., pp. 377–8.
54. Diejomaoh & Anusionwu, 'Education and Income Distribution in Nigeria', pp. 381–2.
55. Terry D. Monson, 'Educational Returns in the Ivory Coast', *Journal of Developing Areas* 13 (1979): 415–30. The internal rate of return is the rate of interest which, used to discount the benefits resulting from an investment (in this case, in education or training), would reduce the net present value of the investment to zero.
56. Ibid., p. 416.
57. Wages of foreign Africans in formalized employment were reported as averaging only one-half those of Ivorians in 1970: see Joshi *et al.*, *Abidjan* Table 2, p. 12. See also the data for 1971 and 1974 in *Ivory Coast: the Challenge of Success*, Table 6.3, p. 130.
58. Fajana, 'Income Distribution in the Nigerian Urban Sector', p. 213.
59. *Outline of the Fourth National Development Plan,* p. 6.
60. Ibid., p. 89; also *Second Progress Report on the Third National Development Plan 1975–80* (Lagos: Central Planning Office, 1979), p. 28, and *Guidelines for the Fourth National Development Plan 1981–85* (Lagos: Federal Ministry of National Planning, 1979), pp. 63–4.
61. Legal minima have long existed in the French-speaking countries. A minimum wage was legislated in Ghana in 1960 and in Nigeria in 1981; in both cases, the purpose was less to erect a floor below which wages might not fall than to concede a general increase in wages in formalized employment.
62. Fajana, 'Income Distribution in the Nigerian Urban Sector', pp. 221–2.
63. Tayo Fashoyin, *Industrial Relations in Nigeria* (London: Longman, 1980), p. 113.
64. *Report of the Commission on the Review of Wages, Salary and Conditions of Service of the Junior Employees of the Governments of the Federation and in Private Establishments, 1963–64* (Lagos: Federal Ministry of Information, 1964), p. 20; *First Report of the Wages and Salaries Review Commission* (Lagos: Federal Ministry of Information, 1970), pp. 11 and 14.
65. By 1979, the highest salary in the Federal public service was 15.5 times that of the lowest (ILO, *First Things First*, Table 43, p. 216), compared with 17.6 times in 1975.
66. John F. Weeks, 'The Impact of Economic Conditions and Institutional Forces on Urban Wages in Ghana', *Nigerian Journal of Economic and Social Studies* 13 (1971), Table IVB, p. 331.
67. ILO, *First Things First*, Table 49, p. 224.

Chapter 4: External Trade

1. *Accelerated Development in Sub-Saharan Africa* op. cit., pp. 19–21 and Statistical Annex, Table 7.

2. Breakdowns of developing countries' exports by principal commodities according to Standard International Trade Classification headings are given for 1965, 1970, 1975 and later years in the *Yearbooks of International Trade Statistics*, vol. I (New York: UN), 1977, Special Table K, and 1980, Special Table N.
3. The annual dollar values of each country's exports and imports have been taken from UNCTAD, *Handbook of International Trade and Development Statistics, Supplement 1980* (New York: UN, 1980), Tables 1.1 and 1.2.
4. Annual statistics showing the direction of trade for each country appear in the *Direction of Trade Yearbooks* (Washington DC: International Monetary Fund). Since 1977, annual statistics showing the five principal importing and exporting partners of each developing country have appeared in the United Nations *Statistical Yearbook* and the November or December issues of the UN *Monthly Bulletin of Statistics.*
5. Elliot Berg, *The Recent Economic Evolution of the Sahel* (Ann Arbor, Center for Research on Economic Development, University of Michigan, 1975), p. 4.
6. *Economic Survey of Africa*, vol. I (Addis Ababa: UN, 1966), pp. 84–6.
7. A. A. Ayida, 'Contractor Finance and Supplier Credit in Economic Growth', *Nigerian Journal of Economic and Social Studies* 7 (1965): 175–88; J. V. Simpson, 'Development Finance: a comment on "Contractor Finance" in Sierra Leone', *Journal of Development Studies* 3 (1967): 175–87; Sayre P. Schatz, 'Crude Private Neo-Imperialism: a new pattern in Africa', *Journal of Modern African Studies* 7 (1969): 677–88; Leslie E. Grayson, 'The Role of Suppliers' Credits in the Industrialization of Ghana', *Economic Development and Cultural Change* 21 (1973): 477–99; Andrzej Krassowski, *Development and the Debt Trap: Economic Planning and External Borrowing in Ghana* (London: Croom Helm, 1974), pp. 80–8.
8. Killick, *Development Economics in Action,* p. 101.
9. Krassowski, *Development and the Debt Trap*, pp. 80, 92.
10. R. J. Bhatia, G. Szapary and B. Quinn, 'Stabilization Program in Sierra Leone', *IMF Staff Papers* 16 (1969): 504–29.
11. *National Development Plan Progress Report 1964* (Lagos: Federal Ministry of Economic Development, 1965), pp. 18–21; Ayida, 'Contractor Finance and Supplier Credit in Economic Growth'.
12. R. U. McLaughlin, 'The Liberian Budget Crisis of 1962/63', in *South of the Sahara: development in African economies* (London: Macmillan, 1972), ed. Sayre P. Schatz.
13. William I. Jones, *Planning and Economic Policy: Socialist Mali and her Neighbours* (Washington DC: Three Continents Press, 1976), pp. 343–4.
14. Manfred Reichardt, 'Stabilizing an Economy – Mali', *Finance and Development* 4 (1967): 281–9.
15. Liberia has also been included under this heading, on the ground that falling rubber and iron ore prices in the early 1960s reduced the profitability of the plantations and mines, although the volumes of exports continued to increase: see McLaughlin, 'The Liberian Budget Crisis of 1962/63', and R. L. Curry, 'Liberia's External Debts and their servicing', *Journal of Modern African Studies* 10 (1972): 621–6.
16. Elliot J. Berg, 'Socialism and Economic Development in Tropical Africa', *Quarterly Journal of Economics* 78 (1964): 558; Krassowski, *Development and the Debt Trap*, pp. 47–9, 72–8; Edwin Dean, *Plan Implementation in Nigeria:*

1962–1966 (Ibadan: Oxford University Press, 1972), pp. 116–24.

17. On the francophone countries' dependence on French support, and its influence on political decisions at the end of the colonial period, see Elliot J. Berg, 'The Economic Basis of Political Choice in French West Africa', *American Political Science Review* 54 (1960): 390–405.
18. The West African Currency Board had been instituted in 1912 with the simple function of issuing and redeeming the currency of The Gambia, the Gold Coast, Nigeria and Sierra Leone against sterling at a fixed parity; it could not extend credit and had no power of monetary control. The replacement of Currency Board money by national currencies issued by national central banks began in Ghana in 1958, in Nigeria in 1959 and in Sierra Leone in 1964.
19. 'Financial Arrangements of Countries using the CFA Franc', *International Monetary Fund Staff Papers* 16 (1969): 289–389.
20. Where crude oil is processed by the same companies that mine it, its value is a matter of negotiation between those companies and the government seeking to tax the mining activity. This negotiated value, used as a basis for tax assessment, is known as the posted price. In 1973, the member-governments of OPEC increased their posted prices unilaterally by very large amounts, taking advantage of fast-growing American demand for imported oil.
21. Douglas Rimmer, 'Development in Nigeria: an Overview', in Bienen and Diejomaoh (eds), op. cit., p. 57.
22. Because of government participation in the ownership of oil producing concessions in Nigeria, some part of the oil extracted has belonged to, and been sold by, a government oil corporation ever since 1971. In 1974, the share of this 'participation crude' in total output became 55 per cent, and since 1979 it has been over 70 per cent. The export price of the Nigerian National Petroleum Corporation has therefore become of more importance, in determining Nigerian earnings from oil, than the posted price of the 'equity crude' remaining with the oil producing companies, although the two prices are closely related. By the end of the 1970s, over 95 per cent of the gross value of Nigerian oil production accrued to the Nigerian state as taxation and earnings of the NNPC.
23. *Ivory Coast: the Challenge of Success* op. cit., pp. 62–8.
24. Ibid., pp. 74–7. Rates of return on educational spending were held down by high failure rates, resulting from a liberal admissions policy combined with maintenance of high examining standards. See Terry D. Monson, 'Educational Returns in the Ivory Coast', op. cit., where it is reported (p. 417) that 60 per cent of enrollees in secondary schools failed to complete the first four-year cycle, 80 per cent failed to complete the full seven years, and 65 per cent of university enrollees failed to complete the four-year programme. In public housing, provided mainly in Abidjan, standards were so high that tenants could not afford the rents unless subsidized.
25. *Ivory Coast: the Challenge of Success*, p. 86.
26. Reported in *West Africa*, 23 June 1980, p. 1101.
27. See the ratios of external public debt to GNP in *World Development Report 1981* (Washington DC: the World Bank, 1981), Annex Table 15, and the debt service ratios in ibid., Annex Table 13.
28. On the experience of Mali, Mauritania, Niger, Senegal and Upper Volta in this period, see Berg, *The Recent Economic Evolution of the Sahel*.
29. According to the estimates in UNCTAD, *Handbook of International Trade and*

Development Statistics, Supplement 1980, Table 7.2.

30. The possession of the Spanish Sahara awarded to Morocco and Mauritania by the departing colonial power in 1976 was contested by the independence movement known as Polisario. Mauritania made peace with Polisario and withdrew from the conflict in August 1979.
31. R. W. Johnson, 'Guinea', in *West African States: Failure and Promise*, ed. John Dunn (Cambridge: Cambridge University Press, 1978), p. 48.
32. The cedi had been devalued by 44 per cent in December 1971, but this decision was partially rescinded by the military government following the second Ghanaian coup in January 1972.
33. SDRs, or Special Drawing Rights, were created by the IMF as a means of increasing international liquidity and first issued in 1970. The basis of their valuation was changed from gold to a basket of sixteen leading currencies in 1974.
34. As in Monique P. Garrity, 'The 1969 Franc Devaluation and the Ivory Coast Economy', *Journal of Modern African Studies* 10 (1972): 627–33.
35. Exports of oil are not deterred, because their value consists very largely in economic rent. Exports of other minerals and of timber are deterred less than exports of agricultural crops, because they have higher ratios of foreign exchange costs (though they might be handicapped by shortage of imported inputs, as has been the case in Ghana).
36. Hopkins, *Economic History of West Africa*, pp. 264–5.
37. As in Reginald H. Green and Ann Seidman, *Unity or Poverty? The Economics of Pan-Africanism* (Harmondsworth: Penguin, 1968).
38. J. Viner, *The Customs Union Issue* (New York: Carnegie Endowment for International Peace, 1950). On the relevance to Africa of theories of customs unions, see Peter Robson, *Economic Integration in Africa* (London: George Allen & Unwin, 1968), and the Introduction to *African Integration and Disintegration: case studies in Economic and Political Union* (London: Oxford University Press, 1967), ed. Arthur Hazlewood.
39. Trade among the member-countries of a free trade area is duty-free so far as concerns their own products, but there is no common external tariff, such as would be found in a customs union (and such as was to be created in the EEC itself).
40. The GATT has been since 1947 the principal agreement governing international trade. As part of a programme intended gradually to eliminate trade barriers, it enunciated non-discrimination as a key principle. The EEC members were among its adherents.
41. Outside West Africa, the invitation was also taken up by Kenya, Tanzania, Uganda, Morocco and Tunisia.
42. The chief negotiator, Dr Pius Okigbo, made forty journeys between Lagos and Brussels: P. N. C. Okigbo, *Africa and the Common Market* (London: Longman, 1967), p. vii.
43. The United Nations Conference on Trade and Development, first convened in 1964, had pressed for preferences to be generally extended to the exports of developing countries in the markets of economically advanced countries, and a chapter had been added to the GATT in 1966 to allow this breaching of the principle of non-discrimination. Preferences for certain manufactured exports of developing countries were initiated in Western Europe and the USA in 1970.

44. The forty-six ACP countries comprised the nineteen associates under the Yaoundé Conventions (former African dependencies of France, Belgium and Italy); twelve former British dependencies in Africa, including The Gambia, Ghana, Nigeria and Sierra Leone; six former British dependencies in the Caribbean, and three in the Pacific; and six other African countries with 'comparable' economies (Ethiopia, Equatorial Guinea, Guinea, Guinea-Bissau, Liberia and Sudan).
45. The Common Agricultural Policy is intended to ensure the living standards of agricultural workers by a combination of variable import levies and of official purchasing of domestic production to support domestic prices.
46. The EUA, or European Unit of Account, is based on the external values of the currencies of EEC members. In 1975, it was worth about £0.42 and $0.93; in 1980, about £0.66 and $1.54.
47. The Nigerian government had disclaimed any right to draw on the European Development Fund.
48. *EEC and the Third World: a Survey 1* (London: Hodder & Stoughton, 1981), ed. Christopher Stevens, Statistical Appx, Tables 9 and 10.
49. Allowing for the rapid inflation between 1975 and 1980, the real value of the aid offered by the second Lomé Convention was about 10 per cent less, on an annual basis, than that offered by the first: see Adrian Hewitt and Christopher Stevens, 'The Second Lomé Convention', in ibid., pp. 50–3.
50. The 550 million EUA allocated to Stabex was intended to be used in five equal annual instalments, but there was also provision for the advance use in any year of a maximum of one-fifth of the following year's instalment. Repayment of transfers from the first Stabex fund brought the total amount available for 1980 to 138 million EUA.
51. World trade in textiles is quantitatively restricted by the Multifibre Arrangement, the third version of which was agreed in December 1981. Under this arrangement, ACP countries are free of the restrictions imposed on the textile exports of other developing countries to the EEC. In West Africa, the Ivory Coast is the country most likely to benefit by this exemption, but it may be doubted that its exports would continue to be tolerated if they were believed seriously to threaten European producer interests. The Ivory Coast also has an interest in the world trade in sugar, having lately developed exporting capacity in this product. In 1982, it was arguing for a quota under the Sugar Protocol of the Lomé Convention, under which the EEC undertakes to import specific quantities of sugar from ACP countries at prices related to those paid for sugar produced in the EEC itself.
52. All West African states together received 43 per cent of these disbursements. The calculations exclude disbursements made for multinational projects. The figures are given in *E.E.C. and the Third World: a Survey 2* (London: Hodder & Stoughton, 1982), ed. Christopher Stevens, Statistical Appx, Table 11.
53. Lynn K. Mytelka, 'A Genealogy of Francophone West and Equatorial African Regional Organisations', *Journal of Modern African Studies* 12 (1974), pp. 300, 305–6.
54. The shadowy Ghana-Guinea Union of 1958 (joined by Mali in 1961) had no economic substance except Ghanaian loans. The *Conseil de l'Entente* was formed by the Ivory Coast in 1959 to detach Dahomey and Upper Volta from Malian influence. Examples of functional or technical unions are various organizations of Senegal River, Chad Basin and Niger River states initiated in

1963–4. For a listing of the West African inter-state organizations extant in the late 1970s, see the Appendix in John P. Renninger, *Multinational Cooperation for Development in West Africa* (New York: Pergamon Press, 1979).

55. 'When Gowon visited Benin in January 1975 in the last rounds of canvassing for WAEC he introduced a Nigerian phenomenon called "spraying" into diplomacy; he literally wrote checks on the spot for every cause' (Olatunde J. B. Ojo, 'Nigeria and the Formation of ECOWAS', *International Organization* 34 (1980), p. 593). Also heavily sprayed were Togo, Guinea and Niger; only the last joined the CEAO.
56. See James O'Connell, 'Political Integration: the Nigerian Case', in *African Integration and Disintegration*, ed. Hazlewood, and Rimmer, 'Development in Nigeria: an Overview'.
57. Further complications are presented by the Mano River Union, a customs union agreement made by Liberia and Sierra Leone in 1973 and joined by Guinea in 1980, and by the Senegambian Confederation agreed by The Gambia and Senegal in 1981, one of whose objectives is an economic union of the two countries.
58. According to the Executive Secretary of ECOWAS: see *West Africa*, 7 June 1982, p. 1493.
59. R. I. Onwuka, 'The ECOWAS Protocol on the Free Movement of Persons: a Threat to Nigerian Security?', *African Affairs* 81 (1982): 193–206.

Chapter 5: Policy Instruments

1. *Third National Development Plan 1975–80* (Lagos: Central Planning Office, special launching edn, 1975), vol. I, p. 27.
2. *Accelerated Development in Sub-Saharan Africa*, op. cit., Statistical Annex, Table 39.
3. *Ivory Coast: the Challenge of Success*, op. cit., Statistical Appx, Table SA30.
4. P. T. Bauer, *West African Trade* (Cambridge: Cambridge University Press, 1954), ch. 19.
5. See, for example, comments in the *Economic Journal* by Polly Hill (63 (1953), 468–71), Peter Ady (63 (1953), 594–607) and B. M. Niculescu (64 (1954), 730–43) on P. T. Bauer and F. W. Paish, 'The Reduction of Fluctuations in the Income of Primary Producers', ibid., 62 (1952), 750–80; John H. Adler, 'The Economic Development of Nigeria: Comment', *Journal of Political Economy* 64 (1956), 425–34; Reginald H. Green, 'Multi-Purpose Economic Institutions in Africa', *Journal of Modern African Studies* 1 (1963), 163–84; G. K. Helleiner, 'The Fiscal Role of the Marketing Boards in Nigerian Economic Development', *Economic Journal* 74 (1964), 582–610; and T. Balogh, *The Economics of Poverty* (London: Weidenfeld & Nicolson, 1966), pp. 67, 124, 250–1.
6. Even in principle, a marketing board could not stabilize income among seasons for individual producers, unless their marketed supplies happened to vary in the same proportions as aggregate supplies varied.
7. *The Tax System of Liberia: Report of the Tax Mission* (New York: Columbia University Press, 1970), pp. 176–9. The Liberian Produce Marketing Corporation was set up as a joint venture of the Liberian government and the East Asiatic Company of Denmark with a monopoly of the export of cocoa, coffee and palm kernels.

8. The United Ghana Farmers' Cooperative Council was the sole licensed buying agent of the Cocoa Marketing Board between 1961 and 1966. On the consequences for cocoa producers of this monopolization of the agency of a monopoly, see Björn Beckman, *Organising the Farmers: Cocoa Politics and National Development in Ghana* (Uppsala: Scandinavian Institute of African Studies, 1976), especially ch. V, and *Report of the Committee of Enquiry on the Local Purchasing of Cocoa* (Accra: Ministry of Information, 1966).
9. The grant element is a measure of the concessionality or softness of a loan. It depends on the difference between the terms of an ODA loan and market terms (with the market rate of interest conventionally taken to be 10 per cent). The present value at the market rate of interest of each repayment of the loan is calculated. The excess of the loan's face value over the sum of these present values, expressed as a percentage of the face value, is the grant element in the loan. The nearer it approaches 100 per cent, the softer the loan.
10. The grant equivalent of an ODA loan is its face value multiplied by its grant element.
11. Aid given in kind, such as technical assistance and food, cannot be spent at the discretion of the recipient; it is embodied in what the donor is supplying. Financial aid may also be tied to the products (in general or in particular) of the donor country. The value attributed to such aid includes a subsidy element in favour of industries or personnel in the donor country.
12. These comparisons are not meant to imply that aid is spent only on capital formation or on imports; it is also used largely for factor payments and transfers abroad arising from technical assistance, for travel and educational expenditures abroad, and possibly for debt servicing.
13. Milton Friedman, 'Foreign Economic Aid: Means and Objectives', *Yale Review* 47 (1958), 24–38; Bauer, *Dissent on Development*, pp. 95–135.
14. *National Development Plan 1962–68* (Lagos: Federal Ministry of Economic Development, 1962), p. 3.
15. Quoted in Tony Killick, 'Commodity Agreements as International Aid', *Westminster Bank Review*, February 1967, p. 19.
16. The integrated programme referred to ten (or possibly eighteen) stockable commodities for which it was hoped to conclude international price-regulating agreements, with finance for buffer stocks being drawn from a Common Fund. From the vast literature on the New International Economic Order, Lars Anell and Birgitta Nygren, *The Developing Countries and the World Economic Order* (London: Frances Pinter, 1980), may be selected.
17. The transactions of the international buffer stock have been overshadowed by the US government's strategic stockpile of tin; see G. W. Smith and G. R. Schink, 'The International Tin Agreement: a Reassessment', *Economic Journal* 86 (1976): 715–28.
18. The Stevenson scheme, under which production in Malaya and Ceylon was restricted in the 1920s, encouraged both the creation of natural rubber-producing capacity elsewhere and the research leading to the production of synthetic rubbers.
19. The inaccuracy of budget forecasting in developing countries is emphasized by N. Caiden and A. Wildavsky in *Planning and Budgeting in Poor Countries* (New York: Wiley, 1974). Ghana was among the countries whose experience they studied.

20. See *Government Finance Statistics Yearbook*, vol. 4, 1980 (Washington DC: International Monetary Fund), pp. 21–3. The functional analyses for African countries are reproduced, with the addition of estimates for Guinea-Bissau and the Ivory Coast, in the World Bank report on *Accelerated Development in Sub-Saharan Africa*, Statistical Annex, Table 41.
21. In 1981 this proportion appears to have been increased to about 40 per cent. Legislation provided that from 1982 it should be 44 per cent.
22. Fuller coverage of military expenditure is given in *Accelerated Development in Sub-Saharan Africa*, Statistical Annex, Table 43, where the primary source of information is the US Agency for International Development. Military expenditure is shown as a ratio of the GNP estimate in 1978. This ratio lies between zero and about 1 per cent in The Gambia, Ghana, Liberia, Niger and Sierra Leone; between 2 and 3 per cent in Benin, the Ivory Coast, Senegal and Togo; between 3 and 4 per cent in Mali and Upper Volta; over 4 per cent in Nigeria; and over 7 per cent in Guinea-Bissau and Mauritania.
23. Loan disbursements and equity purchases by government made for purposes other than management of government liquidity and measured net of amortizations.
24. See Anthony Kirk-Greene and Douglas Rimmer, *Nigeria since 1970: a Political and Economic Outline* (London: Hodder & Stoughton, 1981), pp. 125–7.
25. René Dumont, *False Start in Africa* (London: André Deutsch, 1966), pp. 78–81.
26. On the Nigerian experience, see Edwin Dean, *Plan Implementation in Nigeria: 1962–1966* (Ibadan: Oxford University Press, 1972), ch. 5, and Kirk-Greene and Rimmer, op. cit., pp. 142–5.
27. Heavy taxation of cocoa production in Ghana was on occasion justified by the argument that smaller export supplies would earn larger aggregate proceeds, such was the country's importance in world supply. The argument was equivalent to saying that Ghana alone should bear the costs of a commodity control scheme from which other producing countries could benefit as well as or more than Ghana.
28. Because of the high administrative costs involved in securing payment of rebates to which a claimant is entitled – see A. O. Phillips, 'Nigerian Industrial Tax Incentives: Import Duties Relief and the Approved User Scheme', *Nigerian Journal of Economic and Social Studies* 9 (1967): 315–27.
29. *Marketing Price Policy and Storage of Food Grains in the Sahel: a Survey*, vol. I (Ann Arbor: University of Michigan, Center for Research on Economic Development, 1977).
30. See, for example, Susan Hickok and Clive S. Gray, 'Capital-Market Controls and Credit Rationing in Mali and Senegal', *Journal of Modern African Studies* 19 (1981): 57–73. For a wider discussion, see Sergio Pereira Leite, 'Interest Rate Policies in West Africa', *IMF Staff Papers* 29 (1982): 48–76. In Ghana, real lending rates were negative throughout the period 1976–80, and reached minus 100 per cent in 1977.
31. It might be expected that industrial operations would still be regulated for the protection of health and the environment, and licensing would appear unavoidable in the exploitation of mineral deposits and some other natural resources.
32. For official recognition of these practical difficulties of securing authorization in Nigeria, see *Third National Development Plan*, vol. I, p. 150, and *Fourth National Development Plan 1981–85* (Lagos: National Planning Office, 1981),

vol. I, p. 141. In highly disturbed political conditions, such as obtained in Ghana in 1982, applications for establishment may be deterred simply by the absence of anyone who can make credible commitments on behalf of the government.

33. This expression is used in Robert H. Bates, *Markets and States in Tropical Africa: the Political Basis of Agricultural Policies* (Berkeley & Los Angeles: University of California Press, 1981), ch. 6.
34. See the critical comments on Nigerian LBAs in *Nigeria: Options for Long-Term Development* (Baltimore & London: Johns Hopkins University Press for the World Bank, 1974), pp. 80, 131.
35. See the articles by Eddie Momoh on the Sierra Leone Produce Marketing Board, and by Elizabeth Ohene and Ben Ephson Jr on the Ghanaian Cocoa Marketing Board, in *West Africa*, 9 August 1982, pp. 2038–40: 16 August 1982, pp. 2103–5; and 27 September 1982, pp. 2503–5. In 1982 the Ghanaian CMB was reported to have over 100,000 employees, in spite of which a labour force of students had to be mustered to evacuate the crop to the ports.
36. As a reaction to this abuse in 1982, cocoa farmers in Ghana were to be paid only in non-negotiable cheques, i.e. in marketing board chits instead of buying agents' chits.
37. Although empirical studies are, understandably, few, it has long been recognized that large volumes of export crops are moved illegally among West African countries – e.g. cocoa from Ghana to the Ivory Coast and Togo, cocoa from Nigeria to Benin, groundnuts from Senegal to The Gambia, coffee from Guinea to the Ivory Coast, and all export crops from Sierra Leone to Liberia. For one detailed study, see John Davison Collins, 'The Clandestine Movement of Groundnuts across the Niger-Nigerian Boundary', *Canadian Journal of African Studies* 10 (1976): 259–78. For an attempt to estimate the volume of smuggled cocoa exports from Ghana, see Ashok Kumar, 'Smuggling in Ghana: its Magnitude and Economic Effects', *Nigerian Journal of Economic and Social Studies* 15 (1973): 285–303.
38. The tendency for subsidized commodities to leak into neighbouring countries was discovered at an early stage of independence in Guinea: see Elliot J. Berg, 'Socialism and Economic Development in Tropical Africa', *Quarterly Journal of Economics* 78 (1964): 556–60. For experience in the Sahel in the early 1970s, see the same author's *The Recent Economic Evolution of the Sahel* (Ann Arbor: University of Michigan, Center for Research on Economic Development, 1975), ch. 5.
39. In Guinea, a state monopoly intended to embrace both foreign and domestic trade had been created by 1960. The Ghana National Trading Corporation, established in 1961, is an example of substantial public participation in import trading; in the early 1960s it became the sole import licensee for 'essential' goods. Sale of imported rice at subsidized prices was the responsibility, in Sierra Leone, of a government Rice Corporation, but in 1979 this task was vested in the Sierra Leone Produce Marketing Board, thus producing a neat institutional symmetry (as is observed by Bates, *Markets and States in Tropical Africa*, Appx A): the same organization that collected tax from export agriculture subsidized imported food. Procurement and distribution of agricultural inputs is largely the responsibility of government departments and agencies in all West African countries except the Ivory Coast and Liberia: see *Accelerated Development in Sub-Saharan Africa*, Statistical Annex, Table 32. Important programmes of

public housing were launched in the 1970s in Nigeria (by the Federal Housing Authority) and in the Ivory Coast.

40. See, for example, the accounts of Nigerian experience in Sayre P. Schatz, *Economics, Politics and Administration in Government Lending: the Regional Loans Boards of Nigeria* (Ibadan: Oxford University Press, 1970), and in *Nigerian Capitalism* (Berkeley & Los Angeles: University of California Press, 1977), ch. 12.
41. This analysis of the effects of financial repression is based on Ronald I. McKinnon, 'Financial Policies', in *Policies for Industrial Progress in Developing Countries*, ed. John Cody, Helen Hughes and David Wall (New York: Oxford University Press for the World Bank, 1980), pp. 93–120.
42. *West Africa*, 20 April 1981, p. 894; 18 May 1981, p. 1091; 15 June 1981, p. 1364.
43. Tony Killick, 'Price Controls in Africa: the Ghanaian Experience', *Journal of Modern African Studies* 11 (1973): 405–26.
44. In Ghana, during the period of government by the Armed Forces Revolutionary Council in 1979, traders were flogged and market places demolished in an effort to secure correspondence between market prices and official prices.

Chapter 6: Development Policies

1. 'Modern schools in the field of political science and theory of government . . . assume, indeed, that a government can work on an essentially neutral plane and promote the interests of the community at large . . . Experience in Western countries supports such an assessment of the ability of modern administrations to carry out programmes of economic development which are devised with a view to promoting the welfare of the whole population . . .' (Alfred Bonné, *Studies in Economic Development* (London: Routledge & Kegan Paul, 1957), p. 259). On the persistence of this view, usually held implicitly, among economists and especially economic planners, see Tony Killick, 'The Possibilities of Development Planning', *Oxford Economic Papers* 28 (1976).
2. For impressions of these new perceptions in the development community, see Cambridge University Overseas Study Committee, *Prospects for Employment Opportunities in the Nineteen Seventies*, ed. Ronald Robinson and Peter Johnston (London : HMSO, 1971). For an account of the subsequent evolution of development objectives, see F. A. N. Lisk, 'Conventional Development Strategies and Basic-Needs Fulfilment: a Reassessment of Objectives and Policies', *International Labour Review* 115 (1977).
3. 'Only in a very special sense can we speak of a nation's policy or policies. In general declared policies are nothing but verbalizations of group interests and attitudes . . . Nobody has attained political maturity who does not understand that policy is politics': J. A. Schumpeter, quoted by Arthur Smithies in his memorial of Schumpeter, *American Economic Review* 40 (1950).
4. See Douglas Rimmer, ' "Basic Needs" and the Origins of the Development Ethos', *Journal of Developing Areas* 15 (1981).
5. On the launching of universal primary education in southern Nigeria in the 1950s, see David B. Abernethy, *The Political Dilemma of Popular Education: an African Case* (Stanford: Stanford University Press, 1969).
6. On AOF, see Elliot J. Berg, 'French West Africa', in *Labor and Economic Development* (New York: John Wiley & Sons, 1959), ed. Walter Galenson; on Nigeria, T. M. Yesufu, *An Introduction to Industrial Relations in Nigeria*

(London: Oxford University Press, 1962), ch. VIII, and Robin Cohen, *Labour and Politics in Nigeria* (London: Heinemann, 1974), pp. 197–209.

7. W. F. Stolper, 'The Main Features of the 1962–1968 National Plan', *Nigerian Journal of Economic and Social Studies* 4 (1962); 'Economic Development in Nigeria', *Journal of Economic History* 23 (1963); *Planning without Facts: Lessons in Resource Allocation from Nigeria's Development* (Cambridge, Mass.: Harvard University Press, 1966); 'Social Factors in Economic Planning with special reference to Nigeria' in *Growth and Development of the Nigerian Economy* (East Lansing: Michigan State University Press, 1970), ed. Carl K. Eicher and Carl Liedholm.
8. Economic profitability would take indirect effects into account and might be calculated on the basis of values other than market prices. It was therefore not the same as financial profitability. But the presumption that financial profitability generally understated economic profitability was not accepted; in Nigeria the reverse had sometimes been found.
9. *Seven-Year Plan for National Reconstruction and Development, Financial Years 1963/64–1969/70* (Accra: Planning Commission, 1964), p. 21.
10. A. Akene Ayida, 'Development Objectives', in *Reconstruction and Development in Nigeria* (Ibadan: Oxford University Press, 1971), ed. Ayida and H. M. A. Onitiri.
11. Robert W. Clower, George Dalton, Mitchell Harwitz and A. A. Walters, *Growth without Development: an Economic Survey of Liberia* (Evanston, Illinois: Northwestern University Press, 1966).
12. Samir Amin, *Le développement du capitalisme en Côte d'Ivoire* (Paris: Editions de Minuit, 1967).
13. It is an unfortunate distinction, because the same terms – growth and development – have been used (notably by Schumpeter) in a more illuminating way in theorizing about economic change: 'growth' meaning the accumulation of productive factors of unchanging or slowly changing capacities, to which the economic system can adapt by infinitesimally small steps, and 'development' the making of new productive combinations, or innovations, which disrupt the circular flow of economic life.
14. Unless, of course, poverty and basic needs are defined in such a way that they can be relieved or satisfied only by equality – i.e. unless a deprived person is defined as anyone whose income, or consumption of necessaries, falls short of the mean.
15. Publications which both illustrate and helped to determine this evolution of ideas include the review of evidence by David Turnham and Ingelies Jaeger of *The Employment Problem in Less Developed Countries* (Paris: Development Centre of the OECD, 1971); the reports of ILO Employment Missions including *Towards Full Employment: a programme for Columbia* (Geneva: ILO, 1970), *Matching Employment Opportunities and Expectations: a programme of action for Ceylon* (Geneva: ILO, 2 vols, 1971), and *Employment, Incomes and Equality: a strategy for increasing productive employment in Kenya* (Geneva: ILO, 1972); the article 'What are we trying to measure?' *Journal of Development Studies* 8 (1972), by Dudley Seers, leader of the first two Employment Missions; the United Nations Committee for Development Planning's *Attack on Mass Poverty and Unemployment* (New York: UN, 1972); the first biennial review of *The International Development Strategy* (New York: UN, 1973) by the Secretary-General of the United Nations; the joint study of

the World Bank's Development Research Center and the Institute of Development Studies at the University of Sussex, published as Hollis Chenery *et al.*, *Redistribution with Growth: policies to improve income distribution in developing countries in the context of economic growth* (London: Oxford University Press, 1974); and *Employment, Growth and Basic Needs: a one-world problem* (Geneva: ILO, 1976), the report of the Director-General of the ILO to the Tripartite World Conference on Employment, Income Distribution and Social Progress and the International Division of Labour.

16. See Gavin Kitching, *Development and Underdevelopment in Historical Perspective: Populism, Nationalism and Industrialization* (London: Methuen, 1982).
17. See the discussion in ibid., ch. 4, of the ILO Employment Mission Reports and of Michael Lipton, *Why Poor People Stay Poor: a study of urban bias in world development* (London: Temple Smith, 1977), and E. F. Schumacher, *Small is Beautiful: a study of economics as if people mattered* (London: Blond & Briggs, 1973).
18. On settlement schemes, see K. D. S. Baldwin, *The Niger Agricultural Project* (Cambridge, Mass.: Harvard University Press, 1957), W. Roider, *Farm Settlement Schemes for Socio-Economic Development: the Western Nigerian case* (Munich: Weltforum Verlag, 1971), and R. Chambers, *Settlement Schemes in Tropical Africa* (London: Routledge & Kegan Paul, 1969); on state farms in Ghana, Killick, *Development Economics in Action*, chs. 8 and 9; and on government plantations, R. G. Saylor and C. K. Eicher, 'Plantations in Nigeria: lessons for West African development', in *Change in Agriculture* (London: Duckworth, 1970), ed. A. H. Bunting.
19. See the seminal article by Keith Hart, 'Small scale entrepreneurs in Ghana and development planning', *Journal of Development Studies* 6 (1970).
20. On Ghana, see Killick, *Development Economics in Action*, ch. 7.
21. Douglas Rimmer, 'The Economic Imprint of Colonialism and Domestic Food Supplies in British Tropical Africa', in *Imperialism, Colonialism, and Hunger: East and Central Africa* (Lexington, Mass.: Lexington Books, 1983), ed. Robert I. Rotberg.
22. Douglas Rimmer, 'Development in Nigeria: an Overview', in Bienen and Diejomaoh (eds), op. cit., pp. 37–40, 47–8, 59–62.
23. It is arguable that some of the larger of these ethnic groups owed their identity to the competition for welfare.
24. Received opinion, shaped by H. W. Singer and Raul Prebisch, was that the terms of trade of developing countries were undergoing secular deterioration (*Relative Prices of Exports and Imports of Under-developed Countries* (New York: UN, 1949); Singer, 'The Distribution of Gains between Investing and Borrowing Countries', *American Economic Review (Papers and Proceedings)* 40 (1950); *The Economic Development of Latin America and its Principal Problems* (New York: UN, 1950)). This thesis was embraced by the first UNCTAD in 1964 and has continued to find adherents, at least on the level of economic diplomacy.
25. The principal item in this literature is Eugene Staley's *World Economic Development* (Montreal: ILO, Studies and Reports, series B, no. 36, 1944).
26. R. F. Harrod, 'An Essay in Dynamic Economics', *Economic Journal* 49 (1939); E. D. Domar, 'Capital Expansion, Rate of Growth and Employment',

Econometrica 14 (1946).

27. This line of thought, which was useful in promoting the case for aid in the United States Congress, was suggested by W. W. Rostow, 'The Take-Off into Self-Sustained Growth', *Economic Journal* 66 (1956).
28. Albert O. Hirschman, *Essays in Trespassing: economics to politics and beyond* (Cambridge: Cambridge University Press, 1981), p. 61.
29. A. M. Kamarck, 'Capital and Investment in Developing Countries', *Finance and Development* 8, 2 (1971), p. 2.
30. Kuznets, *Modern Economic Growth*, pp. 80–1, where references are also given to similar findings by other scholars.
31. Investment originally meant outlays expected to bring a direct monetary return to the spender. The term had been extended in the 1930s to include (and perhaps to dignify) expenditures believed to be indirectly gainful, mainly through creating wage-employment. In the 1960s it was further extended to include educational and other social overhead expenditures held to be indispensable for economic growth.
32. Expenditures on education (formal and informal), health, migration and (in some conditions) food and shelter were cited as illustrations of human capital formation in T. W. Schultz's influential address to the American Economic Association in 1960 ('Investment in Human Capital', *American Economic Review* 51 (1961)).
33. *Investment in Education: the report of the Commission on Post-School Certificate and Higher Education in Nigeria* (Lagos: Federal Ministry of Education, 1960), p. 53.
34. Frederick Harbison, 'The African University and Human Resource Development', *Journal of Modern African Studies* 3 (1965), p. 56. The intermediate category comprised persons in occupations normally requiring one to three years of education beyond the O-level stage in secondary schooling, while the high-level category required a university degree or its equivalent.
35. The leading references are P. N. Rosenstein-Rodan, 'Problems of Industrialization of Eastern and South-Eastern Europe', *Economic Journal* 53 (1943); Tibor Scitovsky, 'Two Concepts of External Economies', *Journal of Political Economy* 62 (1954); Ragnar Nurkse, *Problems of Capital Formation in Underdeveloped Countries* (Oxford: Basil Blackwell, 1958).
36. *A Study of Contemporary Ghana*, vol. I, ed. Birmingham, Neustadt and Omaboe, p. 441.
37. *Planning without Facts*, pp. 6–16.
38. *A Study of Contemporary Ghana*, vol. I, pp. 459–61; Edwin Dean, *Plan Implementation in Nigeria: 1962–1966* (Ibadan: Oxford University Press, 1972), pp. 49–63.
39. Douglas Rimmer, 'The Abstraction from Politics: a critique of economic theory and design with reference to West Africa', *Journal of Development Studies* 5 (1969); Killick, 'Possibilities of Development Planning'.
40. Abdelmalek Ben-Amor and Frederick Clairmonte, 'Planning in Africa', *Journal of Modern African Studies* 3 (1965), pp. 494–5.
41. Dudley Seers, 'The Prevalence of Pseudo-Planning', in *The Crisis in Planning* (London: Chatto & Windus, 1972), ed. M. Faber and D. Seers, vol. I, p. 19.
42. Killick, *Development Economics in Action*, pp. 140, 143.
43. Killick, 'Possibilities of Development Planning', p. 162.

44. *An International Economy* (London: Routledge & Kegan Paul, 1956), p. 201; *Economic Theory and Under-Developed Regions* (London: Duckworth, 1957), p. 82.
45. P. T. Bauer, 'The Investment Fetish', in *Equality, the Third World and Economic Delusion* (London: Weidenfeld & Nicolson, 1981).
46. Nicholas Kaldor, 'Taxation for Economic Development', *Journal of Modern African Studies* I (1963), p. 7.
47. Gerald K. Helleiner, 'The Fiscal Role of the Marketing Boards in Nigerian Economic Development, 1947–61', *Economic Journal* 74 (1964), pp. 601–3. Revealingly, Helleiner added: 'It is probable that even if the total economic returns from such items as universities and manufacturing plants were zero, the Governments would, in the present climate of opinion, have regarded them as inherently desirable.'
48. Stolper, *Planning without Facts*, pp. 98–103. It was suggested that the growth of GDP in the 1950s had been partly due to the multiplier effects of investment which were now becoming exhausted.
49. Killick, *Development Economics in Action*, pp. 67–9. The average capital-output ratio (ratio of the capital stock to GNP) is estimated to have risen from 1.78 in 1960 to 3.57 in 1967; in other words, twice as much capital was required in the latter year to produce the output of the former year.
50. Ibid., p. 83.
51. Naseem Ahmad, *Deficit Financing, Inflation and Capital Formation: the Ghanaian experience 1960–65* (Munich: Weltforum Verlag, 1970), pp. 111–18.
52. W. Arthur Lewis, 'Economic Development with Unlimited Supplies of Labour', *The Manchester School* 22 (1954).
53. See J. A. Schumpeter's *Theory of Economic Development* (Cambridge, Mass.: Harvard University Press, 1934), and, on the question of its relevance to developing countries, Douglas Rimmer, 'Schumpeter and the Underdeveloped Countries', *Quarterly Journal of Economics* 75 (1961).
54. These features are characteristic of aid to nearly all parts of the Third World; see Desmond McNeill, *The Contradictions of Foreign Aid* (London: Croom Helm, 1981).
55. Sayre P. Schatz's study of the Federal Loans Board in Nigeria is particularly revealing since the operations of this body were not heavily influenced by political considerations; see *Development Bank Lending in Nigeria* (Ibadan: Oxford University Press, 1964), and 'The Capital Shortage Illusion: government lending in Nigeria', *Oxford Economic Papers* 17 (1965).
56. On Ghana, see Killick, *Development Economics in Action*, especially the evidence summarized on pp. 171–2 and the estimate reported on p. 182 that utilization of non-agricultural capacity fell by one-half during the 1960s. Utilization appears never to have risen above about 40 per cent in the Ghanaian manufacturing sector on even a modest definition of capacity, i.e. single-shift working.
57. James Cobbe, 'Employment and Capital Utilisation: the time aspect', *Journal of Modern African Studies* 19 (1981).
58. On the political manipulation of contract procedures and awards in construction in Nigeria, see Sayre P. Schatz, *Nigerian Capitalism* (Berkeley & Los Angeles: University of California Press, 1977), pp. 170–97. Schatz's material relates to the period before the civil war, but contracting has probably been no less politicized

since that time.

59. This interpretation of the effects of capitalist penetration of pre-capitalist societies was adumbrated in the revision of Marxist teaching made necessary by the Russian revolution of 1917; see Bill Warren, *Imperialism: pioneer of capitalism* (London: New Left Books, 1981), ch. 4. Paul A. Baran, 'On the Political Economy of Backwardness', *The Manchester School* 20 (1952), illustrates the ideological position reached after the Second World War. For applications to Africa, see Samir Amin, *Neo-Colonialism in West Africa* (Harmondsworth: Penguin Books, 1973) and Walter Rodney, *How Europe Underdeveloped Africa* (London: Bogle-L'Ouverture, 1972). There is considerable overlapping between the neo-Marxist perspective and Latin American 'dependency theory'; see, for instance, André Gunder Frank, 'The Development of Underdevelopment', *Monthly Review* 18 (1966), and Theotonio dos Santos, 'The Structure of Dependence', *American Economic Review* 60 (1970).
60. Thus Robert Szereszewski, *Structural Changes in the Economy of Ghana 1891–1911* (London: Weidenfeld & Nicolson, 1966), estimated that national income per head in the Gold Coast had roughly quadrupled between 1890 and the late 1950s, while population had increased in the same period by about the same factor. The Ivory Coast is the leading West African case of rapid economic growth in an alleged neo-colony.
61. See Warren, *Imperialism*, pp. 39–44.
62. See Peter Kilby, 'Manufacturing in Colonial Africa', in *Colonialism in Africa 1870–1960*, vol. 4 (Cambridge: Cambridge University Press, 1975), ed. Peter Duignan and L. H. Gann.
63. There had been complaints in the 1930s of structural distortion in the economies of Germany, Italy and Japan; these professedly 'have-not nations', like their postwar successors in the Third World, cited persistent shortage of foreign exchange as evidence of their economic malformation. See Douglas Rimmer, 'Have-Not Nations: the Prototype', *Economic Development and Cultural Change* 27 (1979).
64. Singer, 'Distribution of Gains between Investing and Borrowing Countries', pp. 476–7.
65. John Spraos, 'The Statistical Debate on the Net Barter Terms of Trade between Primary Commodities and Manufactures', *Economic Journal* 90 (1980).
66. Kilby, *Industrialization in an Open Economy*, ch. 4; A. N. Hakam, 'The Motivation to Invest and the Locational Pattern of Foreign Private Industrial Investments in Nigeria', *Nigerian Journal of Economic and Social Studies* 8 (1966); *Ivory Coast: the Challenge of Success* (Baltimore & London: Johns Hopkins University Press for the World Bank, 1978), pp. 230–1.
67. Kwame Nkrumah, *Africa Must Unite* (1963), quoted in Killick, *Development Economics in Action*, p. 185.
68. William F. Steel, 'Import Substitution and Excess Capacity in Ghana', *Oxford Economic Papers* 24 (1972).
69. Killick, *Development Economics in Action*, pp. 187–8.
70. *Ivory Coast: the Challenge of Success*, pp. 242–4.
71. *Nigeria: Options for Long-Term Development* (Baltimore & London: Johns Hopkins University Press for the World Bank, 1974), p. 83. The overall EPC for Ivorian manufacturing in 1975, according to the World Bank study cited in the text, was 1.42.

72. A notable exposition was Ian Little, Tibor Scitovsky and Maurice Scott, *Industry and Trade in some Developing Countries* (London: Oxford University Press for the Development Centre of the OECD, 1970).
73. W. B. Morgan, 'Food Imports of West Africa', *Economic Geography* 39 (1963).
74. This suggestion is made for Nigeria in Francis Teal, 'The Supply of Agricultural Output in Nigeria, 1950–1974', *Journal of Development Studies* 19 (1983).
75. Scott R. Pearson, J. Dirck Stryker, Charles P. Humphreys *et al.*, *Rice in West Africa: policy and economics* (Stanford: Stanford University Press, 1981). The RCR differs from the DRC coefficient only in expressing the opportunity costs of domestic factors directly in terms of foreign exchange.
76. Carl K. Eicher and Doyle C. Baker, *Research on Agricultural Development in Sub-Saharan Africa: a critical survey* (East Lansing: Department of Agricultural Economics, Michigan State University, 1982), p. 121.
77. See Killick, *Development Economics in Action*, ch. 9.
78. *Ivory Coast: the Challenge of Success*, pp. 220–1.
79. President Houphouët-Boigny, reported in *West Africa*, 13 October 1980, p. 2001.
80. This timing is also partly explained by the lagged reaction of output to earnings in cocoa. New plantings take several years to come into bearing.
81. The producer price was reduced from 72 shillings per headload to 60s in 1959, as a so-called voluntary contribution of the farmers to the financing of the second development plan, which promised establishment in the next five years of not less than 600 factories. There were further reductions to 54s in 1961 and 40s in 1965. The fall in the producer price during this period of Ghana's big push toward industrialization would be even greater if allowance were made for the rising cost of living.
82. The average of 189,000 tonnes for 1978–80 shown in Table 28 is derived from the *Annual Reports* of the Central Bank of Nigeria. The provisional figures for these years reported in the *Cocoa Market Report* (March 1983) of the Gill and Duffus Group yield an average of only 144,000 tonnes.
83. In 1973 the Nigerian federal government announced that the producer prices of crops controlled by the marketing boards would be fixed in future with no trading surpluses in view, and the taxes on export crops were removed in the following year. In 1977 it was acknowledged that, cocoa apart, the Nigerian marketing boards had become only residual buyers of crops because of the growth of the home market.
84. See Amin, *Le développement du capitalisme en Côte d'Ivoire*, and R. H. Green, 'Reflections on Economic Strategy, Structure, Implementation and Necessity', in *Ghana and the Ivory Coast: perspectives on modernisation* (Chicago: University of Chicago Press, 1971), ed. Philip Foster and A. R. Zolberg.
85. See M. O'Connor, 'Guinea and the Ivory Coast: contrasts in economic development', *Journal of Modern African Studies* 10 (1972), and R. W. Johnson, 'Guinea', in *West African States: failure and promise* (Cambridge: Cambridge University Press, 1978), ed. John Dunn.
86. In Ghana, legislation since 1968 has reserved many small-scale industrial and service activities to Ghanaians. In Nigeria, the indigenization decree of 1977, carrying further restrictions introduced in 1972, grouped enterprises in three categories: those reserved to Nigerians, those in which a minimum Nigerian shareholding of 60 per cent was required, and the remainder in which a minimum

Nigerian shareholding of 40 per cent was required. Alongside (and, in Nigeria, overlapping with) these requirements of private indigenous control or participation has been public participation in foreign enterprises deemed to be of strategic importance (in, for example, mining, banking and insurance).

87. Robert H. Bates, *Markets and States in Tropical Africa: the political basis of agricultural policies* (Berkeley: University of California Press, 1981), pp. 95, 122, 131–2. On the size-distribution of landholdings, see Eddy Lee, 'Export-Led Rural Development: the Ivory Coast', *Development and Change* 11 (1980).
88. See Kitching, *Development and Underdevelopment in Historical Perspective*, pp. 136, 180, and his quotation (p. 32) from Marx on the populists – 'They all want the impossible, namely the conditions of bourgeois existence without the necessary consequences of those conditions.'

Chapter 7: Conclusions

1. The official rate of 2.75 Ghanaian cedis to the dollar was retained in April 1983, but foreign exchange outlays were to attract surcharges at 750 or 990 per cent, while earnings would be increased by bonuses paid at the same rates. The system has the advantages of avoiding the stigma of an outright devaluation, and of retaining bureaucratic control of foreign exchange transactions and hence the opportunities of corruption. In Sierra Leone, the official rate was retained in January 1983, but holders of foreign exchange could convert it into leones at nearly three times the official rate, and the proceeds of these transactions were allocated by auction among prospective importers; the parallel market was thus being legitimized, but not integrated with the official market.
2. Because of the glut in export markets, the following changes occurred:

	Oil production (million barrels per day)	*Av. NNPC sales price ($ per barrel)*	*Govt revenues from oil ($ billion)**
1980	2.052	35.20	24.0
1981	1.433	38.77	16.5
1982	1.300*	35.65*	14.5

(* approximations.)

3. In Liberia, the True Whig government of President W. R. Tolbert was overthrown by a group of disgruntled army NCOs. In spite of the execution of several politicians and much revolutionary rhetoric, it subsequently became plain that little had changed in the conduct of Liberian affairs apart from an increase in military prerogatives. In Ghana, junior officers and NCOs seized power briefly in 1979, before the restoration of civilian government in that year, establishing the vengeful regime of the Armed Forces Revolutionary Council under the leadership of Flight-Lieutenant J. J. Rawlings. Following a New Year's Eve party in 1981, soldiers overthrew the government of the People's National Party headed by President H. Limann, and set up the People's National Defence Council under the chairmanship of J. J. Rawlings. Though much torn by internal dissension, and desperately ambivalent in its economic policies, this soldiers' government retained power in 1982–3 with the support of urban wage-earners, setting up popular authorities on the Libyan model and lowering still further the economic efficiency of their country.

4. Since 1970, centrally collected revenues have been allocated among the constituent States of the Nigerian federation on the criteria principally of equality and of population size according to the 1963 census results. From 1981, 44 per cent of centrally collected revenues were distributed among the States (and the local government councils they contained) – about 36 per cent on the basis of equality among States, 36 per cent according to population, and 28 per cent according to other criteria (primary school enrolments, internal revenue effort, and derivation).
5. The so-called 'inverted-U hypothesis' – that income inequality first rises and later falls in a country as its economy grows – was suggested by Simon Kuznets ('Economic Growth and Income Inequality', *American Economic Review* 45 (1955)), and has been extensively, though not invariably, supported by historical studies.
6. The argument is in John Kenneth Galbraith, *The Nature of Mass Poverty* (Cambridge, Mass.: Harvard University Press, 1979), ch. 6.
7. On the long-standing conservative argument that schooling in Africa ought to be appropriate to current circumstances or vocationally relevant, see P. J. Foster, 'The Vocational School Fallacy in Development Planning', in *Education and Economic Development* (New York: Aldine, 1966), ed. C. A. Anderson and M. S. Bowman.
8. Helleiner's index of the commodity terms of trade of Nigeria (1953=100) averaged 63 in 1936–8, 65 in 1946–8 and 108 in 1954–6 (G. K. Helleiner, *Peasant Agriculture, Government and Economic Growth in Nigeria* (Homewood, Ill.: Richard D. Irwin, 1966), Appx, Table IV-A-6, p. 500); the official index of the Ghanaian terms (1954=100) averaged 55 in 1948–50 and 85 in 1954–6 (*1963 Statistical Year Book* (Accra: Central Bureau of Statistics, 1966), Table 131).
9. For explanation of equity and participation crude, see Chapter 4, notes 20 and 22.
10. 'The growth of agricultural exports has been the main element carrying the economy,' wrote Arthur Lewis in 1966 (*Reflections on Nigeria's Economic Growth* (Paris: Development Centre of the OECD, 1967), p. 16).
11. According to R. W. Johnson, 'it is impossible to over-estimate the importance of bauxite which provides the entire basis for the country's (and the regime's) economic survival' ('Guinea', in *West African States* (Cambridge: Cambridge University Press, 1978), ed. John Dunn, p. 48).
12. As in the celebrated UN experts' report on *Measures for the Economic Development of Under-Developed Countries* (New York: United Nations, 1951), p. 16.
13. See Robert H. Bates, *Markets and States in Tropical Africa* (Berkeley: University of California Press, 1981), ch. 7.
14. The World Bank estimates that consumption per head of purchased energy increased by a factor of about 3 in Nigeria and the Ivory Coast between 1960 and 1979, and about 2.5 in Ghana. In Liberia, Mauritania, Niger, Togo and Upper Volta, the increase appears as more than fivefold; only in Benin and Guinea was there less than a doubling. (*Accelerated Development in Sub-Saharan Africa* (Washington DC: World Bank, 1981), Statistical Annex, Table 6).
15. Douglas Rimmer, 'Schumpeter and the Underdeveloped Countries', *Quarterly Journal of Economics* 75 (1961).
16. In particular, the colonial administrations, like the governments that succeeded them, found it difficult to accept the rationality of abstaining from food production.

BIBLIOGRAPHICAL NOTE

Detailed references to the literature relevant to West African economies have been made in the text of this work. The objects of the present note are only to indicate principal sources of information and to suggest further reading on the subject. Most of the estimates of economic quantities originate in national statistical offices, central banks, planning ministries and other official institutions. These estimates are generally unreliable, as has been frequently observed in the text, but a work of the character of the present book would be impossible without making use of them. The best one can do is to use the figures with due sensitivity to their likely margins of error (which implies not using them for purposes in which accuracy is required). In Ghana and Nigeria, the immediacy, and possibly also the quality, of economic statistics have deteriorated since about the late 1960s.

For most readers, the official estimates will be most accessible in publications of the international agencies where they have also sometimes been refined. Particularly convenient are the statistical annex of the *World Development Reports* of the World Bank, published annually since 1978, and the *International Financial Statistics* of the IMF, which appear both monthly and as a yearbook. The IMF also publishes yearbooks on the *Direction of Trade* and *Government Finance Statistics*. Useful UN serials include the *Monthly Bulletin of Statistics,* the *Statistical Yearbook*, the *Yearbook of International Trade Statistics* and, especially, the *Demographic Yearbook*. A *Year Book of Labour Statistics* is published by the ILO, but the information it gives on West Africa is inevitably sparse. The UNCTAD *Handbook of International Trade and Development Statistics* is especially useful for its estimates of the terms of trade. Educational statistics are to be found in the UNESCO *Statistical Yearbook*, and aid statistics in publications of the Development Assistance Committee of the OECD – the annual report on *Development Cooperation*, and the *Geographical Distribution of Financial Flows to Developing Countries*. The dependence of tabulated information in this work on these sources will be obvious.

Among direct West African sources, publications of the central banks tend to be more readily obtainable and up-to-date than those of other official institutions. The annual reports of the Central Bank of Nigeria and the Bank of Sierra Leone are valuable and still widely distributed. The *Banque Centrale des Etats de l'Afrique de l'Ouest* (embracing Benin, the Ivory Coast, Niger, Senegal, Togo and Upper Volta) produces both an annual report and an excellent monthly dossier of *Notes d'information et statistiques*. Useful information can be extracted from national development plans, but it is important to remember that economic projections are not the same thing as (and can usually be expected to differ widely from) realized results – a point which should also be made of budget statements.

The World Bank report on *Accelerated Development in Sub-Saharan Africa* (1981) reproduces estimates for African countries from the statistical annex of the *World Development Report 1981*, and surveys the constraints on economic growth in Africa, the character of the development policies that have been followed, and the results achieved. It was the work of a team headed by Elliot J. Berg, some of whose earlier economic writings on Africa have been cited in the text. A similar performance by a member of the World Bank staff, Shankar Acharya, is 'Perspectives and Problems of Development in Low Income Sub-Saharan Africa', *World Development* 9 (1981); this too can be strongly recommended, though the author's slant is toward East rather than West Africa. The reports of World Bank missions have been published on Nigeria (1955 and 1974), Senegal (1975) and the Ivory Coast (1978); only the last (*Ivory Coast: the Challenge of Success*) retains much currency. The Bank sponsored jointly with the OECD the study of *Migration in West Africa* (1981), by K. C. Zachariah and Julien Condé, which deals with nine countries of the region (not including Nigeria) and was referred to in Chapters 1 and 3 of this work.

The *Surveys of African Economies* published by the IMF cover the francophone West African countries other than Guinea and Mali in volume 3 (1970), the anglophone countries in volume 6 (1975), and Mali in volume 7 (1977). These surveys give information for short time periods and make little attempt at interpretation; consequently they are of very limited use.

The concern of the ILO for the unemployed, underemployed and unequal, and for unsatisfied basic needs, led to a report on *Total Involvement: a Strategy for Development in Liberia* (1974); studies of urban development and employment in *Abidjan* (by Heather Joshi *et al.*, 1976) and *Lagos* (by O. J. Fapohunda *et al.*, 1978); and reports by the Jobs and Skills Programme for Africa on *First Things First: Meeting the Basic Needs of the People of Nigeria* (1981) and *Ensuring Equitable Growth: a Strategy for Increasing Employment, Equity and Basic Needs Satisfaction in Sierra Leone* (1981).

Scholarly economic work on West Africa was more abundant before about 1970 than it has been since. The oustanding recent work is Tony Killick's *Development Economics in Action* (1978), which deals with economic policies in Ghana in the period 1961–72 but has a much wider relevance. Killick was also principal author, with Robert Szereszewski, of the now very dated volume I of *A Study of Contemporary Ghana* (1966), edited by W. B. Birmingham *et al.*, while volume 2 (1967) consists largely in the demographic studies of J. C. Caldwell. Many illusions have been lost in Ghana since the early 1960s, including some by the present writer; for one intellectual conversion, see Richard Jeffries, 'Rawlings and the Political Economy of Underdevelopment in Ghana', *African Affairs* 81 (1982).

Several important economic studies of Nigeria were published in the

1960s, including Gerald K. Helleiner's *Peasant Agriculture, Government and Economic Growth in Nigeria* (1966), Wolfgang F. Stolper's *Planning without Facts* (1967), Adebayo Adedeji's *Nigerian Federal Finance* (1969), and Peter Kilby's *Industrialization in an Open Economy: Nigeria 1945–1966* (1969). The output since the civil war has been less distinguished, but Sayre P. Schatz's *Nigerian Capitalism* (1978), although based on field work conducted mostly in the early 1960s, still gives much illumination, and the collection of studies on *The Political Economy of Income Distribution in Nigeria* (1981), edited by Henry Bienen and V. P. Diejomaoh, can be recommended. Anthony Kirk-Greene and Douglas Rimmer, *Nigeria since 1970* (1981), outlines the political and economic evolution of the country in the period 1970–80.

The Ivory Coast has been poorly served by economists writing in English (though useful papers by Eddy Lee and T. D. Monson have been cited in the text). Samir Amin's influential and much-cited *Le développement du capitalisme en Côte d'Ivoire* (1967) has not been translated. The report of the World Bank mission to the country in 1975 has already been mentioned. On Sierra Leone, the only substantial recent work is John Levi *et al.*, *African Agriculture: Economic Action and Reaction in Sierra Leone* (1976). On Liberia, little of significance since the controversial *Growth without Development: an Economic Survey of Liberia* (1966), by R. W. Clower *et al.* On the Sahelian countries, there are important studies financed by US AID and directed by Elliot Berg at the Center for Research on Economic Development, University of Michigan – *The Recent Economic Evolution of the Sahel* (1975) and the two-volume survey of *Marketing, Price Policy and Storage of Food Grains in the Sahel* (1977).

Recent books on agricultural issues include Keith Hart, *The Political Economy of West African Agriculture* (1982), and the much more technical *Rice in West Africa: Policy and Economics* (1981), by Scott R. Pearson *et al.*, which is concerned with the Ivory Coast, Liberia, Mali, Senegal and Sierra Leone.

Among works by non-economists, powerful stimuli can be obtained from Stanislav Andreski, *The African Predicament: a Study in the Pathology of Modernisation* (1968), and Robert H. Bates, *Markets and States in Tropical Africa: the Political Basis of Agricultural Policies* (1981). In *West African States: Failure and Promise* (1978), edited by John Dunn, the papers on Guinea by R. W. Johnson and Liberia by Christopher Clapham are particularly instructive.

The best source of current information is probably still the weekly *West Africa*, which has been published continuously in London since 1917.

A highly selective (and inexpensive) short-list of further reading on the matters dealt with in this book would be the Berg report on *Accelerated Development in Sub-Saharan Africa*, Killick's *Development Economics in Action*, and Bates's *Markets and States in Tropical Africa.*

INDEX

Index